12

Band 21 · Dezember 1985

isotopenpraxis

AKADEMIE-VERLAG

ISSN 0021-1915 Isotopenpraxis, Berlin 21 (1985) 12, 409–456

EVP 9,— M

An unsere Autoren

Die Zeitschrift veröffentlicht Beiträge zu folgenden Themen: Kontrolle und Analyse radioaktiver und stabiler Isotope, Anwendung radioaktiver und stabiler Isotope in allen Gebieten der Forschung und Technik, der Industrie, Landwirtschaft und Medizin, Herstellung radioaktiver und stabiler Isotope sowie von markierten Verbindungen, kernelektronische Geräte für die Isotopenanwendung (Dickenmesser, Dichtemesser u. ä.) Detektoren für isotopentechnische Arbeiten und den Strahlenschutz in Isotopenlaboratorien, Meßtechnik und Strahlenschutz, Dosimetrie in Isotopenlaboratorien, kerntechnische Geräte für die Isotopenanwendung (Beton, Bestrahlungsanlagen u. ä.), Strahlenchemie, Radiochemie, ökonomische und perspektivische Fragen, Fragen der Weiterbildung.

1. Aufnahme von Manuskripten

Zur Veröffentlichung angenommen werden nur druckreife, noch nicht anderweitig publizierte Beiträge in deutscher, englischer oder russischer Sprache zum Themenbereich der Zeitschrift und zwar

Originalarbeiten
Kurze Mitteilungen
Aus der Praxis
Zusammenfassende Berichte
Tagungsberichte
Informationen.

Sind der Beitrag oder Teile daraus bereits veröffentlicht oder ist eine anderweitige Veröffentlichung vorgesehen, so muß dies der Redaktion mit genauer Angabe der Zweitveröffentlichung mitgeteilt werden. Mit der Annahme des Manuskriptes behält sich der Verlag das Recht der Vervielfältigung, Verbreitung und Übersetzung innerhalb der gesetzlichen Schutzfrist vor. Auszugsweiser Nachdruck oder photomechanische u. a. Wiedergabe sind an eine schriftliche Genehmigung des Verlages gebunden. Manuskripte, Zeichnungen und Photos von veröffentlichten Arbeiten werden nur auf den besonderen und bis zum Zeitpunkt der Veröffentlichung mitgeteilten Wunsch des Autors zurückgesandt.

Bei der unverlangten Einreichung von Manuskripten wird empfohlen, eine Befürwortung eines der Herausgeber oder Redaktionsausschußmitglieder beizufügen.

Manuskripte sind einzureichen bei der Redaktion, 1086 Berlin, Leipziger Str. 3–4.

2. Form der Manuskripte

Die Autoren werden gebeten:

2 Manuskripte, eines davon in Original-Maschinenschrift, einseitig und mit 2zeiligem Abstand einzureichen; gut leserliche Photokopien sind zulässig; auf dem 1. Blatt (unter Weglassung der Bezeichnung)

den Titel der Arbeit
Vorname-Initialen und Namen sämtlicher Autoren
gegebenenfalls die Institution, in der sie tätig sind
eine kurze Zusammenfassung der Arbeit von höchstens 15 Zeilen
Keywords in alphabetischer Reihenfolge und
als Fußnote die Anschrift der Institution anzugeben;

wenn möglich, auf gesonderten Blättern Namen, Titel der Arbeit und Zusammenfassung in den beiden anderen obengenannten Sprachen 2fach beizufügen;

Literaturhinweise (fortlaufend numeriert), Fußnoten, Tabellen, Abbildungen (in 2–3facher Veröffentlichungsgröße), Abbildungsunterschriften (getrennt von den Abbildungen mit Zeichenerklärungen, Quellenangaben, falls erforderlich, und gegebenenfalls mit einem Vermerk für besondere Wünsche bezüglich der Abbildungsgröße), private und Dienstanschrift der Verfasser mit akademischen Titeln und ausgeschriebenen Vornamen auf besonderen Blättern ans Ende des Manuskriptes zu stellen;

Wünsche für den Kolumnentitel (Seitenüberschrift) bekanntzugeben;

Absätze durch Einzüge von mindestens 3 Anschlägen kenntlich zu machen;

die Maßeinheiten des SI-Systems („Systeme International") zu verwenden;

Korrekturen im Manuskript unmittelbar an der zu korrigierenden Stelle anzubringen;

Formeln deutlich lesbar und möglichst groß einzutragen;

Farbabbildungen und nachträgliche Änderungen oder Ergänzungen in Abbildungen zu vermeiden;

hervorzuhebende Textstellen zu unterstreichen, und zwar halbfetten Satz mit einer geraden schwarzen Linie, kursiven Satz (Schrägschrift) mit einer blauen Wellenlinie (alle Eigennamen). Sperrungen gestrichelt;

besondere Buchstaben durch farbige Unterstreichung zu kennzeichnen, und zwar Formelzeichen in

Griechisch-Kursiv (α, β, ...) rot,
Kursiv (A, B, ...) blau
halbfetter Kursiv ($\boldsymbol{A}$, $\boldsymbol{B}$, ...) 2 Linien blau (Vektoren),
halbfetter Grotesk ($\mathbf{A}$, $\mathbf{B}$, ...) 2 Linien violett (Tensoren);

bei Literaturangaben auf Vollständigkeit zu achten. In nachstehender Reihenfolge müssen angegeben sein:

a) bei *Zeitschriften*: Vorname-Initialen und Namen sämtlicher Autoren, Titel der Zeitschrift (nach „Periodica Chimica" oder möglichst wenig gekürzt), Bandnummer (unterstrichen), Jahreszahl (in Klammern) Heftnummer (nur bei Zeitschriften, die jedes Heft mit S.1 beginnen) sowie die Seitenzahl, auf der die zitierte Arbeit beginnt.

b) bei *Büchern*: Initialen der Vornamen und Name des Autors (Herausgeber werden durch den Zusatz „ed." gekennzeichnet), voll ausgeschriebener Titel des Buches in der Originalsprache, evtl. Bandnummer, voll ausgeschriebener Name des Verlages, Nummer der Auflage (außer der ersten), Erscheinungsort(e), Erscheinungsjahr (hier nicht in Klammern setzen!) und evtl. Kapitelnummer, Seitenzahl u. ä.

c) bei *Sammelbänden, Tagungsberichten*: Es soll zusätzlich der Titel des zitierten Beitrags angegeben werden. Die Seitenzahl muß unbedingt angegeben werden. Aus dem Zitat muß hervorgehen, ob es sich um einen Tagungssammelband oder um einen noch nicht veröffentlichten Vortrag handelt. Zwischen Tagungsort und Erscheinungsort des entsprechenden Tagungsberichts ist zu unterscheiden.

3. Korrekturen

Der Autor erhält zwei Umbruchabzüge, von denen einer möglichst schnell korrigiert an die Redaktion zurückgereicht werden soll. Auf Wunsch des Autors kann die Korrektur auch von der Redaktion gelesen werden.

Durch nachträgliche Manuskriptänderungen verursachte Korrekturkosten gehen, sofern sie 10% der Satzkosten übersteigen, zu Lasten des Autors.

Der Autor erhält 50 Sonderdrucke kostenlos. Weitere Sonderdrucke können bis zur Rückgabe der Korrekturabzüge auf Kosten des Autors bei der Redaktion bestellt werden.

Bezugsmöglichkeiten

Bestellungen sind zu richten

— in der DDR an den Postzeitungsvertrieb unter Angabe der Kundennummer des Bestellenden oder an den AKADEMIE-VERLAG BERLIN, DDR-1086 Berlin, Leipziger Straße 3–4, PF-Nr. 1233;
— im sozialistischen Ausland an eine Buchhandlung für fremdsprachige Literatur oder an den zuständigen Postzeitungsvertrieb;
— in der BRD und Berlin(West) an eine Buchhandlung oder an die Auslieferungsstelle KUNST UND WISSEN, Erich Bieber OHG, Wilhelmstraße 4–6, D-7000 Stuttgart 1;
— in den übrigen westeuropäischen Ländern an eine Buchhandlung oder an die Auslieferungsstelle KUNST UND WISSEN, Erich Bieber GmbH, Dufourstraße 51, CH-8008 Zürich;
— im übrigen Ausland an den Internationalen Buch- und Zeitschriftenhandel; den Buchexport Volkseigener Außenhandelsbetrieb der Deutschen Demokratischen Republik, DDR-7010 Leipzig, Postfach 160; oder an den AKADEMIE-VERLAG BERLIN, DDR-1086 Berlin, Leipziger Straße 3–4.

Zeitschrift „Isotopenpraxis"

Herausgeber: Zentralinstitut für Isotopen- und Strahlenforschung der Akademie der Wissenschaften der DDR.
Verlag: Akademie-Verlag Berlin, DDR-1086 Berlin, Leipziger Straße 3–4, PF-Nr. 1233; Fernruf: 2236221 oder 2236229; Telex-Nr.: 114420;
Bank: Staatsbank der DDR, Berlin, Kto.-Nr.: 6836-26-20712.
Chefredakteur: Prof. Dr. H. Hübner, Leipzig.
Anschrift der Redaktion der Zeitschrift Isotopenpraxis: Prof. Dr. H. Hübner, DDR-7010 Leipzig, Karl-Marx-Platz, Hauptpostamt, PSF 370; Fernruf: 281018.
Veröffentlicht unter der Lizenznummer 1527 des Presseamtes beim Vorsitzenden des Ministerrates der Deutschen Demokratischen Republik.
Gesamtherstellung: VEB Druckerei „Thomas Müntzer", DDR-5820 Bad Langensalza.
Erscheinungsweise: Die Zeitschrift „Isotopenpraxis" erscheint monatlich. Die 12 Hefte eines Jahrganges bilden einen Band. Bezugspreis je Band 240,— DM zuzüglich Versandspesen; Preis je Heft 20,— DM. Der gültige Jahresbezugspreis für die DDR ist der Postzeitungsliste zu entnehmen.
Bestellnummer dieses Heftes: 1079/21/12.

AN (EDV) 12831
00900

Isotopenpraxis 21 (1985) 12, S. 409 bis 412

Transition State Theory in Isotope and Radiation Chemistry

W. Stiller[1]), *R. Haberlandt*[1]), *Sz. Vass*[2])

([1]) Central Institute of Isotope and Radiation Research of the Academy of Sciences of the GDR, Leipzig; [2]) Central Research Institute for Physics of the Hungarien Academy of Sciences, Budapest)

A short survey of the history of the transition state theory (TST) including new developments is given. In isotope chemistry reaction-kinetic isotope effects can be calculated via TST using the concept of molecular partition functions. Several examples are given. In radiation chemistry TST is applied on particular to ionic reactions. The general procedure is exemplified by ion-molecule reactions.

Es wird ein kurzer Überblick über die Geschichte der Theorie des Übergangszustandes (TST), einschließlich neuer Entwicklungen gegeben. In der Isotopenchemie können kinetische Isotopieeffekte nach der TST mit Hilfe von Zustandssummen berechnet werden. Mehrere Beispiele werden gegeben. In der Strahlenchemie wird die TST insbesondere auf ionische Reaktionen angewendet. Das allgemeine Vorgehen wird am Beispiel der Ion-Molekül-Reaktionen näher erläutert.

Keywords

gases; ions; isotopes; isotope effects; kinetics; radiation chemistry; reaction kinetics

Fifty years ago *Henry Eyring* published his famous paper on "The Activated Complex in Chemical Reactions" [1] ideas of which became a milestone in chemical reaction kinetics. The historical events give us the opportunity to review some details of it was found, developed and applied especially in the field of isotope and radiation chemistry.

Some historical remarks on TST

When we consider a bimolecular reaction between A and B forming products C and D the bimolecular (forward) reaction rate constant k is defined by the following rate equation:

$$-\frac{\mathrm{d}n_A}{\mathrm{d}t} = k n_A n_B \qquad (1)$$

Here n_A and n_B are the concentrations of the reactive components A and B, respectively. In general the "rate constant" $k(T)$ is independent of the n's. It depends more or less sensitively on the temperature T and possibly on further parameters. For this reason the denotation "rate coefficient" instead of rate constant is justified and will be used throughout the paper.

The study of especially the temperature behaviour of k has a long history beginning — at a quantitative level — in the middle of the last century (see e.g. [2]).

Based on experimental work in 1889 *Svante Arrhenius* formulated what we now know as the *Arrhenius* equation [3]:

$$k(T) = A \cdot \exp\,(-E_A/k_B T) \qquad (1)$$

A, preexponential ("frequency") factor; E_A, *Arrhenius* (experimental) activation energy; k_B, *Boltzmann's* constant.

The thermodynamic foundation of rate theory based on eq. (1) was given already by *van't Hoff* [4] in the last century. In the following decades, however, extensive experimental work in kinetics led chemists to the conclusion that the quantity A in eq. (1) is not a constant, but a temperature-dependent quantity in general. Parallel to these findings theorists succeeded in explaining $A(T)$ from a molecular (microscopic) point of view. Thus decisive progress was made by *M. Trautz* and *W. C. McC. Lewis* [5, 6] using the simple kinetic theory of gases. They regarded the reacting molecules A and B as hard spheres with radii r_A and r_B, respectively, which resulted in an effective reaction diameter $d = r_A + r_B$. If the particles move with the relative speed $v = |\boldsymbol{v}_A - \boldsymbol{v}_B|$ the preexponential factor can be derived in the following way:

$$A(T) = \pi d^2 v = \pi d^2 \left(\frac{8 k_B T}{\pi M}\right)^{1/2} \sim T^{1/2} \qquad (2)$$

where M is the reduced mass, $M^{-1} = M_A^{-1} + M_B^{-1}$, of both reacting particles.

When in 1935 *H. Eyring* published his basic paper giving an interesting scheme for calculating $k(T)$ (and also $A(T)$ according to (1)) he could extend and generalize important kinetic work done by *Marcelin* [7] (consideration of potential energy surfaces) and *Herzfeld* and *Smallwood* [8] in the same way as *Pelzer* and *Wigner* [9] (treatment of reaction as a "translational motion" (along a reaction path). A central point of *Eyring's* paper, however, was the concept of an "activated complex". This was supposed to be, so to speak, an intermediate reactive compound state (transition state, TS) at the saddle point of the potential energy surface (see Fig. 1). Furthermore in his transition state theory (TST) *Eyring* assumed the activated complexes to be in an equilibrium state with the original reactants. The supposition is of great practical importance because it opens the possibility of calculating concentrations via the statistical mechanics of equilibrium systems. The chemical process of the reaction itself can be understood as a free translational motion from the reactants to the products along a "reaction path". Mathematically speak-

Addresses: [1]) DDR-7050 Leipzig, Permoserstraße 15; [2]) H 1525 Budapest, P. O. Box 49

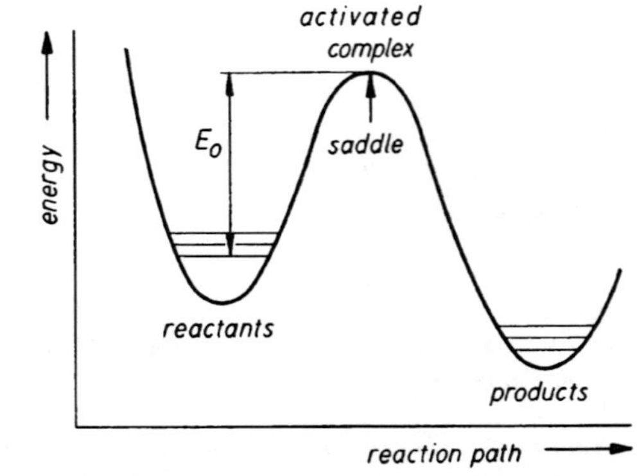

Fig. 1. Transition State Theory (TST) of activated complexes (principle)

ing *Eyring* obtained for the reaction

$$A + B \rightleftharpoons {}^{\ddagger}X \rightarrow C + D \tag{3}$$

the following expression (the symbol $\ddagger$ denotes the TS[3]):

$$k(T) = \varkappa \frac{k_B T}{h} {}^{\ddagger}K e^{-E_0/k_BT}$$

$$= \varkappa \frac{k_B T}{h} \frac{{}^{\ddagger}Q}{Q_A Q_B} e^{-E_0/k_BT}$$

$$= \varkappa \frac{k_B T}{h} e^{\Delta^{\ddagger}S/k_B} e^{-\Delta^{\ddagger}H/k_BT} \tag{4}$$

In eq. (3) ${}^{\ddagger}K$ is a quasi-equilibrium constant. Effects such as tunnelling, steric hindrance or non-adiabatic behaviour can be represented globally in the transmission coefficient $\varkappa$. h is *Planck's* constant. The quantities $Q_A(T)$, $Q_B(T)$, and ${}^{\ddagger}Q$ are the partition functions of the reactants A, B and the activated complex state ${}^{\ddagger}X$, respectively. E_0 is the difference between the zero point energies of ${}^{\ddagger}X$ and A + B.

In its classical version, eq. (4), TST combines ease of calculation with the hard work of finding the experimental data for ${}^{\ddagger}X$ in order to arrive at ${}^{\ddagger}Q$. Tab. 1 summarizes advantages and disadvantages of TST from the present point of view.

Tab. 1. Comparison of some conceptual advantages (+) with more serious deficiencies (−) of conventional TST

+	−
relatively simple expression for calculation $k(T)$	no prescription for calculation $\varkappa$
good idea of how (at least in principle) a reaction could occur	some problems with the assessment of ${}^{\ddagger}Q$ (how to choose the configuration of the TS)
consideration of structural effects (masses, distances, moments of inertia, vibrational frequencies ...)	neglect of essential dynamic effects, limitation of validity to thermal equilibrium[1]) between A, B and ${}^{\ddagger}X$

[1]) For the treatment of non-equilibrium situations see e.g.: *W. Stiller, R. Schmidt*, Z. physik. Chemie (Leipzig) **261** (1980) 271

Since 1935 a great variety of proposals have been made to improve or generalize TST [10, 11]. A first extension to reaction in condensed phases was already given by *Eyring* himself, together with *Wynne-Jones* [12], introducing the concept of activity coefficients into TST. TST was further advanced by *Eliason* and *Hirschfelder* [13] in that they considered dynamical aspects more thoroughly. An idea of such a procedure will be given in an example (ion-molecule reactions) in our radiation-chemical part. Another way was taken by *Marcus* [14], who introduced a system of curvilinear coordinates in order to find separable approximations for the potential function. Finally mention should be of the quantum-mechanical version of TST elaborated by *Miller* [15] and the work by *Truhlar* and coworkers, who combined TST of bimolecular reactions with the criterion of minimum state density (see e.g. [16]).

TST in isotope research

Parallel to the development of theoretical work in reaction kinetics which culminated in *Eyring's* papers in 1935 rapid progress was also made in isotope chemistry at that time. Obviously stimulated by the sensational discovery of an isotope of hydrogen (deuterium) by *Urey, Brickwedde* and *Murphy* [17], above all *Gilbert N. Lewis*, the theorist in chemistry, together with *R. T. Macdonald* started isotope separation at Berkeley in 1932. The experimental work done there from 1932 to World War II has recently been described by *Bigeleisen* [18]. Theoretical aspects of combining isotope research with TST were summarized by *Bigeleisen* and *Wolfsberg* [19] and later by *Vojta* [20].

Let us consider the following competing isotope reactions:

$$A_1 + B_1 \overset{k_1}{\rightleftharpoons} {}^{\ddagger}X_1 \rightarrow C_1 + D_1 \tag{5}$$

$$A_2 + B_2 \overset{k_2}{\rightleftharpoons} {}^{\ddagger}X_2 \rightarrow C_2 + D_2 \tag{6}$$

In eqs. (5), (6) the species A_1 is assumed to contain the lighter isotope (e.g. H), the species A_2 the heavier one (e.g. D).

The (reaction-)kinetic isotope effect (KIE) can be described by the ratio k_1/k_2 (e.g. k_H/k_D). Now we can make use of *Eyring's* TST-formula (4) leading to a good approximation to

$$\frac{k_1}{k_2} = \frac{{}^{\ddagger}Q_1}{{}^{\ddagger}Q_2} \frac{Q_{A_2} \cdot Q_{B_2}}{Q_{A_1} \cdot Q_{B_1}} \cdot e^{-E_0/k_BT} \tag{7}$$

Depending on whether the ratio

k_1/k_2	is	– greater than	1	a regular	KIE
		– equal to	1	we obtain no	
		– less than	1	an inverse	

If in particular the second reactant B_1 or B_2 in the concurrency scheme (5), (6) is independent of isotopic properties ($B_1 = B_2 = B$), the KIE depends only on the ratios of partition functions (PFs) of the species involved:

$$k_1/k_2 = ({}^{\ddagger}Q_1/{}^{\ddagger}Q_2)/(Q_{A_2}/Q_{A_1})\, e^{-E_0/k_BT} \tag{8}$$

For predictions of the KIE the molecular structures of A_1, A_2, ${}^{\ddagger}X_1$, and ${}^{\ddagger}X_2$ have to be taken into account. It should be mentioned, however, that this concept of using PF's of the molecules is only correct for ideal systems, in which the intermolecular interaction energy between the molecules is negligibly small as compared with the total energy. Assuming such a situation the PF Q of a molecule within a system of N particles in the volume V is defined by

$$Q(T, V, N) = \sum_n e^{-E_n/k_BT} \tag{9}$$

Neglecting e.g. intramolecular energy transfer to a good approximation the (total) energetic eigenvalues E_n of the molecule can be split up into a sum of parts for translation, external rotation, internal rotation of molecular groups, vibrations, excitations of electrons and nuclei:

$$E_n = E_{tr} + E_{rot} + E_{rot,i} + E_{vib} + E_{el} + E_{nuc} \tag{10}$$

[3]) Throughout the paper we write the symbol $\ddagger$ before the quantities X, K, Q, S, H and so on.
This is in agreement with the latest IUPAC recommandation published in Pure & Appl. Chem. **54** (1982) 1239.

Putting eq. (10) into (9) the PF can be factorized into

$$Q = Q_{tr} \cdot Q_{rot} \cdot Q_{rot,i} \cdot Q_{vib} \cdot Q_{el} \cdot Q_{nuc} \quad (11)$$

with the well-known results (see e.g. [20—22], e.g. for translation (classical)

$$E_{tr} = \frac{1}{2M} p^2 \qquad Q_{tr} = V\left(\frac{2\pi M k_B T}{h^2}\right)^{3/2} \quad (12)$$

external rotation (classical)

$$\left.\begin{aligned} E_{rot} &= \frac{h^2}{2I} n(n+1) \\ Q_{rot} &= \frac{1}{s} \frac{\pi^{1/2}(8\pi^2 k_B T)^{3/2} (I_a I_b I_c)^{1/2}}{h^3} \end{aligned}\right\} \quad (13)$$

international rotation (quantum-mechanical)

$$E_{rot,i} = \tfrac{1}{2} V_0(1 - \cos r\varphi) \qquad x = V_0/2k_B T$$

$$Q_{rot,i} = \frac{2}{r} \frac{(2\pi I_r k_B T)^{1/2}}{h} e^{-x}\left\{I_0(x) - \frac{u^2}{24} I_1(x)\right\} \quad (14)$$

vibration (quantum-mechanical)

$$E_{vib} = h\nu(n + 1/2) \qquad Q_{vib} = \prod_i \frac{e^{-u_i/2}}{1 - e^{-u_i}} \quad (15)$$

(M, p, mass, momentum of the molecule, respectively; I, I_r, I_a, I_b, I_c, moments of inertia; n, quantum number of rotation or vibration; V_0, maximum of the potential of hindrance of the internal rotation; s and r, number of symmetry of the external and internal rotation, respectively; $u_i = h\nu_i/k_B T$, ν_i, vibronic frequency; $I_0(x)$, $I_1(x)$, modified *Bessel* functions)

According to eq. (8) for the calculation of KIE expressions for the ratios Q_{A_2}/Q_{A_1} and Q_1/Q_2 are needed. In what follows we assume the electronic and nucleonic parts Q_{el} and Q_{nuc} to be equal for A_2 and A_1 and $^{+}X_1$ and $^{+}X_2$.

Then such a ratio of isotopic species with N atoms can be calculated via eqs. (11)—(13) and (15) (neclecting internal rotation effects) in the form:

$$\frac{Q_2}{Q_1} = \left(\frac{M_2}{M_1}\right)^{3/2} \left(\frac{I_{a_2} I_{b_2} I_{c_2}}{I_{a_1} I_{b_1} I_{c_1}}\right)^{1/2} \frac{s_1}{s_2} \prod_i^{3N-6} \frac{e^{-u_{2i}/2}}{e^{-u_{1i}/2}} \frac{1 - e^{-u_{1i}}}{1 - e^{-u_{2i}}} \quad (16)$$

By means of the "product rule" of *Teller* and *Redlich* [23] it was possible to transform "rotational" expressions into vibronic ones via the relation (M and m, molecular and atomic masses, respectively):

$$\left(\frac{M_2}{M_1}\right)^{3/2} \prod_{k=1}^{N} \left(\frac{m_{1k}}{m_{2k}}\right)^{3/2} \left(\frac{I_{a_2} I_{b_2} I_{c_2}}{I_{a_1} I_{b_1} I_{c_1}}\right)^{1/2} = \prod_{i=1}^{3N-6} \frac{u_{2i}}{u_{1i}} \quad (17)$$

By substituting eq. (17) into (16) a convenient reduced ratio f of PS's could be defined by *Bigeleisen* and *Goeppert-Mayer* [24]:

$$f = \frac{Q_2'}{Q_1'} = \prod_{i=1}^{3N-6} \frac{u_{2i}}{u_{1i}} \frac{\sinh (u_{1i}/2)}{\sinh (u_{2i}/2)} \quad (18)$$

Introducing (18) into (8) one is led to the so-called *Bigeleisen* equation for a KIE:

$$\frac{k_1}{k_2} = \frac{^{+}\nu_1}{^{+}\nu_2} F \quad (19)$$

The frequencies $^{+}\nu_1$ and $^{+}\nu_2$ represent the "reactive" vibronic degrees of freedom of the activated complexes $^{+}X_1$ and $^{+}X_2$ resulting in a decay of $^{+}X_1$ and $^{+}X_2$, respectively. F is equal to $f_A/\,^{+}f'$ for $B_1 = B_2 = B$ and equal to $f_A f_B/^{+}f'$ in the more general case $B_1 \neq B_2$. In the quantity f' the summation goes only to $3N$-7 because one vibronic degree of freedom has already been considered by $^{+}\nu$.

From eqs. (18) and (19) it can be concluded: With the help of *Eyring's* TST a KIE can be calculated via reduced ratios of PF's if all the vibrational frequencies of the reactants and the activated complexes are known which should be determined by spectroscopic methods or (theoretically) by normal coordinate analysis.

For practical calculations of KIE the structure of the activated complexes is the main problem. Obviously this point is a consequence of the shortcomings of TST itself. Different proposals have been made to circumvent the flaw. Thus in calculating KIE for the thermal dissociation of deuterated molecules e.g. *Rühlemann* [25] assumed the transition state to have a vanishing frequency of the C--H stretching vibration by setting the corresponding force constant equal to zero. *Johansen* [26] used the BEBO concept (BEBO = binding energy — binding order) originally developed by *Johnston* and *Parr* in order to fix the structure of ^{+}X. He found maximum KIE for symmetric positions of the saddle point.

TST in Radiation Chemistry

Energy-rich radiation leads to numerous ionic processes in radiation chemistry. In irradiated gas phase systems, especially ion-molecule reactions occur. It is interesting to note that as early as 1936 it was again *H. Eyring* [27] (together with *Hirschfelder* and *Taylor*) who did pioneering work along the lines of his 1935 paper. For the reaction $H_2^+ + H_2 \xrightarrow{k} H_3^+ + H$ he obtained the then surprisingly high value of $k = 2{\cdot}07 \cdot 10^{-9}$ cm^3 · molecule · s^{-1}, which is in good agreement with later measurements made by *Stevenson* and *Schissler* [28] and *Klein* and *Friedman* [29]. *Eyring's* result led to a remarkable change in the interpretation of the mechanism acting in radiation chemistry. Before the *Eyring* paper (1936) was published radiation physicists and chemists believed in long-lived ionic clusters. This led to an overestimation of the ionic mechanisms. As a consequence of *Eyring's* paper specialists were convinced that ionic species do not have a great influence on gas phase radiation chemistry, because they react so rapidly even with molecules. Today we know, of course, that ionic, radical and excitation mechanisms determine the events in radiation chemistry.

Now let us come to a very successful combination of *Eyring's* TST of 1935 with a more dynamical "ansatz" formulated by *Eliason* and *Hirschfelder* [13]. The result is the following expression for the bimolecular rate soefficient $k(T)$ studied, e.g., by *Morokuma*, *Eu* and *Karplus* [30]:

$$k(T) = \left(\sum_n e^{-E_n/k_B T} \int_0^\infty k(E, n)\, \varrho(E)\, e^{-E/k_B T} \mathrm{d}E\right) Q_{tr} Q_{vr} \quad (20)$$

(n, set of quantum numbers for the initial vibration-rotation state of the reactants; Q_{tr}, translational part of the PF of the relative motion of reactants A and B per unit volume; Q_{vr}, vibration-rotation part of the PF of A and B)

The function $\varrho(E)$ is defined by

$$\varrho(E) = 4\pi M(2ME)^{1/2}/h^3\,, \qquad E = \frac{M}{2} v^2 \quad (21)$$

Thus the rate coefficient $k(T)$ in eq. (20) is an average of the "rate kernel" $k(E, n)$ (which is unknown) over the equilibrium distributions of E and n. For translationally driven reactions [31] $k(E, n)$ has for $E \geq E_0$ the property:

$$k(E, n) = k(E) \quad (\text{and} \quad = 0\,, \quad \text{for} \quad E \leq E_0) \quad (22)$$

(E_0, threshold energy of reaction)

Equating the kinetic expression of $k(T)$ in eq. (20) with that of *Eyring* the following integral equation for $k(E)$ is obtained ($\varkappa = 1$):

$$\int_{E_0}^{\infty} dE\, k(E)\, \varrho(E)\, e^{-E/k_B T} = \frac{k_B T}{h} \frac{{}^{+}Q_{vr}}{Q_{vr}} e^{-E_0/k_B T} \qquad (23)$$

Due to *Marcus* $k(E)$ can be linked with the energy-dependent cross section $\sigma_R(E)$:

$$k(E) = \frac{{}^{+}W(E - E_0)}{h\varrho(E)} \qquad \sigma_R(E) = \pi\lambda^2\, {}^{+}W(E - E_0) \qquad (24)$$

$\lambda = \hbar/(2\, ME)^{1/2}$, *de Broglie* wave length of the reactants. The TST-expression of the rate coefficient $k(T)$ gives the kernel $k(E)$ and via the quantity ${}^{+}W(E - E_0)$ (number of states of activated complexes with energies $E > E_0$) the microscopic cross section $\sigma_R(E)$. The last quantity is a central measure of reactive binary collision dynamics (molecular beam technique). In this manner *Eyring's* theory enables us — in combination with collision dynamics — to find reactive cross sections of bimolecular reactions, in particular of ion-molecule reactions. *Eyring's* theory was also useful in other fields of radiation-induced reactions, such as unimolecular ionic fragmentation or the decomposition kinetics of neutral excited molecules.

Summary

TST gives a general framework how the process of a chemical reaction can be understood in principle. At the same time TST provides a scheme for assessing rate coefficients of even more complex reactions. The accuracy of the rate coefficients calculated in general depends sensitively on our knowledge of potential energy surfaces. Here the chemists hope that sufficient fast computers will be available in the future for calculating potential energy surfaces at least in particular cases [32, 33].

TST in its conventional form or in one of the modified versions can be applied also to reactions in more sophisticated chemical situations (solvent phenomena, kinetic isotope effects) or reactions under extreme conditions (e.g. high temperatures and pressures, irradiation). In so much TST is an important tool for evaluating rate coefficients of reactions in isotope and radiation chemistry [34].

Received, May 15, 1985

References

[1] *H. Eyring*, J. chem. Phys. **3** (1935) 107
[2] *L. Wilhelmy*, Pogg. Ann. **81** (1850) 422, 499
[3] *S. Arrhenius*, Z. physik. Chem. **4** (1889) 226
[4] *J. H. van't Hoff*, Lectures on Theoretical and Physical Chemistry, part 1: Chemical Dynamics. London: Edward Arnold 1898 (Comp. also *J. H. van't Hoff*, Etudes de Dynamique Chimique. Amsterdam: F. Muller 1884)
[5] *M. Trautz*, Z. physik. Chemie **96** (1916) 1
[6] *W. C. McC. Lewis*, J. chem. Soc. **109** (1916) 796; **111** (1917) 457; **113** (1918) 471
[7] *R. Marcelin*, C. R. Hebd. Seances Acad. Sci. **158** (1914) 116, 407; Ann. Phys. **3** (1915) 120
[8] *K. F. Herzfeld, H. M. Smallwood* in *H. S. Taylor*: Physical Chemistry, Macmillan 1931, p. 151; see also *K. F. Herzfeld*: Kinetische Theorie der Wärme (Müller-Pouillets Lehrbuch der Physik), 1925
[9] *H. Pelzer, E. Wigner*, Z. physik. Chemie, Abt. B **15** (1932) 445
[10] *K. J. Laidler, M. C. King*, J. phys. Chem. **87** (1983) 2657
[11] *A. Kuppermann*, J. phys. Chem. **83** (1979) 171
[12] *W. F. K. Wynne-Jones, H. Eyring*, J. chem. Phys. **3** (1935) 492
[13] *M. A. Eliason, J. O. Hirschfelder*, J. chem. Phys. **30** (1959) 1426
[14] *R. A. Marcus*, J. chem. Phys. **41** (1964) 2614, 2624; **46** (1967) 959
[15] *W. H. Miller*, J. chem. Phys. **61** (1974) 1823; **62** (1975) 1899; **63** (1975) 1166; **65** (1976) 2216
[16] *B. C. Garrett, D. G. Truhlar*, J. chem. Phys. **70** (1979) 1593
[17] *H. C. Urey, F. G. Brickwedde, G. M. Murphy*, Phys. Rev. **39** (1932) 164; **40** (1932) 1
[18] *J. Bigeleisen*, J. chem. Educ. **61** (1984) 108
[19] *J. Bigeleisen, M. Wolfsberg*, Adv. chem. Phys. **1** (1958) 15–76 (75 ref.)
[20] *G. Vojta*, Theorie thermodynamischer Isotopieeffekte. In: Lehrbuch der Kernphysik, Bd. III/*G. Hertz* (ed.), Leipzig: Teubner 1962, S. 89–116 (30 ref.)
[21] *R. Haberlandt, G. Vojta*, Kernenergie **5** (1962) 321
[22] *R. Münze, H. Koch, A. Hermann, G. J. Beyer, R. Haberlandt* (Autorenkollektiv): Radiochemische Methoden (Teil I), Lehrmaterial zum Postgradualstudium „Radiochemie" an der TU Dresden, 1980
[23] *G. Herzberg*, Infrared and Raman Spectra of Polyatomic Molecules, van Nostrand, Inc. Princeton New Jersey, New York 1954
[24] *J. Bigeleisen, M. Goeppert-Mayer*, J. chem. Physics **15** (1947) 261
[25] *S. Rühlemann*, Z. phys. Chemie (Leipzig) **242** (1969) 353
[26] *H. Johansen*, Isotopenpraxis 8 (1972) 407
[27] *H. Eyring, J. O. Hirschfelder, H. S. Taylor*, J. chem. Phys. **4** (1936) 479
[28] *D. P. Stevenson, D. O. Schissler*, J. chem. Phys. **23** (1955) 1353
[29] *F. S. Klein, L. J. Friedman*, J. chem. Phys. **41** (1964) 1789
[30] *K. Morokuma, B. C. Eu, M. Karplus*, J. chem. Phys. **51** (1969) 5193
[31] *W. J. Chesnavich, M. T. Bowers*, J. phys. Chem. **83** (1979) 900
[32] *D. G. Truhlar, W. L. Hase, J. T. Hynes*, J. phys. Chem. **87** (1983) 2664
[33] *P. Pechukas*, Ber. Bunsenges. phys. Chemie **86** (1982) 372; Annu. Rev. phys. Chem. **32** (1981) 159
[34] *L. Melander, W. H. Saunders*, Jr., Reaction Rates of Isotopic Molecules. New York: Wiley 1980

Isotopenpraxis 21 (1985) 12, S. 413 bis 417

Originalarbeiten

Quantum Isotope Effects in Gas Transport Through Polymers

P. Mercea, V. Toşa

(Institute of Isotopic and Molecular Technology)[1])

A diffusing gas molecule of sufficiently light mass (e.g. H_2 and D_2) dispersed in sufficiently small cavities of a polymer will posses ground — state quaentization of energy, both in the cavity and in the transition state. Hence one can observe isotope effects in the activation energies for diffusion and permeation and enthalpies of sorption for two isotopic species, providing the mass ratio of the lighter to the heavier is sufficiently small. These isotope effects were determined for four sorts of polymers and were used to estimate parameters involved in the "Free Volume Theory" of gas diffusion through polymers.

Ein diffundierendes Gas von ausreichend leichter Masse (z. B. H_2 und D_2), das in kleinen Hohlräumen eines Polymeren dispergiert ist, besitzt einen quantisierten Grundzustand der Energie, und zwar sowohl in dem Hohlraum als auch im Übergangszustand. Deshalb kann man Isotopieeffekte in der Aktivierungsenergie der Diffusion und der Permeation beobachten und die Sorptionsenthalpien für die beiden isotopen Spezies bestimmen, vorausgesetzt, das Massenverhältnis des leichteren zum schwereren Isotop ist genügend klein. Diese Isotopieeffekte sind an vier verschiedenen Polymeren bestimmt worden und wurden dazu verwendet, um bestimmte Parameter nach der Theorie des freien Volumens für die Gasdiffusion durch Polymere zu bestimmen.

Keywords

deuterium; diffusion; hydrogen isotopes; isotope effects; permeability; polyethylenes; polymers; pvc;

Introduction

According to the theories of gas transport through polymers the rate controlling step in the diffusion process is a thermally activated jump, of the penetrant gas molecule, over a sufficiently high, well — defined potential energy barrier, separating an initially-penetrant occupied cavity from a neighboring penetrant — free one [1]. These cavities or "vacancies" in the polymer represent regions of relative minimum potential energy in the configurational space. In the classical case for two isotopes the physical properties which determine the barrier height are the same. Thus, for instance the *Lennard—Jones* potential parameters for H_2 and D_2 are identical and the work to overcome the energy barrier depends only on the size and not on the mass of the penetrant molecule. The *Eyring* activation energy for diffusion E_D:

$$E_D^{(i)} = -R \frac{d \ln D}{d(1/T)} \tag{1}$$

is numerically equal to the barrier height and thus it is equal for both isotopes [2].

In the multidimensional configuration space of the diffusion system there are two effective lenghts on which one must focus attention. First it is λ_c, the effective length characterizing the cavity, or the normal state Q_i, and the second is λ_b, characterizing the saddle point, or the activated state Q_i^+ Fig. 1 [3].

When λ_c and λ_b are sufficiently small and the mass m_i of the penetrant molecule is very small there is a ground state quantization of the translational energy of the penetrant inside the cavity (ε_i), and at the saddle point (ε_i^+) [4]. Thus in the case of two small isotopes ($m_1 = H_2$ and $m_2 = D_2$) the rotational and vibrational partition functions in the normal and activated state are equal, but the zero — point translational energies differ. One can assume that because the potential barrier between two neighboring cavities is large enough one can neglect the effects due to direct quantum — mechanical tunneling through the barrier for H_2 and D_2. Therefore ε_i and ε_i^+ are given to a good approximation by the ground — state translational energy of a free particle in two boxes with infinitely high walls [4]:

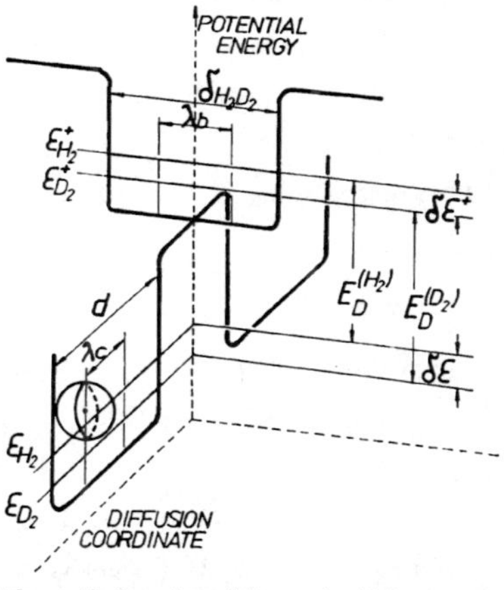

Fig. 1. Schematic drawing of the potential energy surface for H_2 and D_2 diffusion in polymers

$$\varepsilon_i = \frac{h^2 \cdot N}{8 \cdot m_i \cdot \lambda_c^2} \tag{2}$$

$$\varepsilon_i^+ = \frac{h^2 \cdot N}{8 \cdot m_i \cdot \lambda_b^2} \tag{3}$$

[1]) Address: 3400 Cluj — Napoca 5, P. O. Box 700, Romania

where: h and N are *Planck's* constant and *Avogadro's* number. Not too far from the room temperature ($E_d^{(i)} \geq 5\,RT$) the penetrant isotope, H_2 or D_2, is essentially in its ground state. The amount of thermal activation required by the penetrant to surmount the potential energy barrier is thus changed by the amount:

$$E_D^{(i)} = \varepsilon_i^+ - \varepsilon_i \tag{4}$$

Since both ε_i and ε_i^+ decrease with increasing isotopic mass m_i, one can expect to observe an isotope effect in the measured activation energies for diffusion of H_2 and D_2:

$$E_D^{(1)} - E_D^{(2)} = \delta E_D = (\varepsilon_2 - \varepsilon_1) - (\varepsilon_2^+ - \varepsilon_1^+) = \delta\varepsilon^+ - \delta\varepsilon \tag{5}$$

On the other hand the activation energies of permeation, $E_P^{(i)}$ and enthalpies of sorption $H_S^{(i)}$ can be determined from the temperature dependence of the permeability, $P^{(i)}$, respectively of the solubility coefficients, $S^{(i)}$, according to the following relations [5]:

$$E_P^{(i)} = -R\,\frac{\mathrm{d}\ln D}{\mathrm{d}\,(1/T)} \tag{6}$$

$$H_S^{(i)} = -R\,\frac{\mathrm{d}\ln S}{\mathrm{d}\,(1/T)} \tag{7}$$

It is known that $E_P^{(i)}$ and $H_S^{(i)}$ can be derived only from the activated Q_i^+ and respectively from the normal state Q_i [6]. Thus, the difference in the effective activation energy of permeation, δE_P, and enthalpy of sorption, $\delta\Delta H_S$, of isotopes 1 and 2 in the same medium are given by:

$$E_P = E_P^{(1)} - E_P^{(2)} = \varepsilon_1^+ - \varepsilon_2^+ = \delta\varepsilon^+ \tag{8}$$

$$\Delta H_S = \Delta H_S^{(1)} - \Delta H_S^{(2)} = \varepsilon_1 - \varepsilon_2 = \delta\varepsilon \tag{9}$$

From Eq. (5), (8) and (9) one can find:

$$\delta E_P = \delta E_D + \delta\Delta H_S \tag{10}$$

According to (8) and (9) introducing (2) and (3) into (5) one obtains:

$$\delta E_D = \frac{h^2 N}{8}\left(\frac{1}{m_1} - \frac{1}{m_2}\right)\cdot\left(\frac{1}{\lambda_b^2} - \frac{1}{\lambda_c^2}\right) \tag{11}$$

where δE_D may be positive or negative, according to λ_c being smaller or larger than λ_b.

Experimental results

Four sorts of polymer membranes were investigated:

1) polyethylenetherephthalate PET, 2) polyethylene PE, 3) and 4) two sorts of polyvinyl chloride PVC 1 and PVC 2 plastified with 25 respectively 35 parts of diocthylphthalate at 100 parts of weight of polymer. The hydrogen had at least 99,8% purity, while deuterium, of more than 98,5% purity, was obtained from the reaction between D_2O (99,75% purity) and metallic Na in excess. The apparatus and experimental method has been described in a previous paper [7]. The time lag and the steady state diffusion methods were used to determine the diffusion $D^{(i)}$ and permeation $P^{(i)}$ coefficients for H_2 and D_2, in the range of 290 – 340 K [8]. Since in the case of simple gases diffusion through polymers there is very little interaction between the penetrating molecules and the polymer, one may write:

$$P^{(i)} = D^{(i)}\cdot S^{(i)} \tag{12}$$

Hence from the $D^{(i)}$ and $P^{(i)}$ data we also determined, in the same temperature range, the solubility coefficients $S^{(i)}$. According to Eq. (1), (6) and (7) one can calculate for both isotopes H_2 and D_2 the activation energies $E_D^{(i)}$ and $E_P^{(i)}$ and the apparent solvation enthalpies $\Delta H_S^{(i)}$. Therefore, δE_D, δE_P and $\delta\Delta H_S$ can be derived and by means of Eq. (8 – 11) one can evaluate the λ_b and λ_c effective lengths. The obtained results are summarized in Tab. 1.

Tab. 1. Isotope effects in permeation and diffusion activation energies and solvation enthalpies of H_2 and D_2 in polymers. Cavity and barrier parameters of the polymers

Polymer	δE_P	δE_D	$\delta\Delta H_S$	λ_c	λ_b	d	δ	$\bar{v}$
	kcal/mol			Å				Å³
PET	0·3	−0·2	0·5	0·45	0·59	3·38	3·52	20·2
PE	0·2	−0·2	0·4	0·52	0·72	3·45	3·62	21·5
PVC 1	0·7	0·3	0·4	0·50	0·39	3·43	3·32	21·1
PVC 2	0·4	0·2	0·2	0·74	0·51	3·67	3·44	25·9

In Eq. (2) a constant shape factor has been incorporated into λ_c^2. The value of this factor varies from 3 to 1 as one goes from a cubical to a spherical box [4]. Hence, assuming that the vacities which can accomodate H_2 and D_2 are spherical and have approximatively the same volume, an actual estimate of the geometrical cavity diameter d can be made. For this, the effective diameter, σ_{H_2,D_2}, of the H_2 and D_2 molecule in the polymer must be known. A good approximation is to take for σ_{H_2,D_2} the quantum mechanically corrected *Lennard-Jones* potential diameter of 2,928 Å [9]. Then the mean apparent H_2 or D_2 occupied cavity diameter d and the mean cavity volume $\bar{v}$ will be:

$$d \cong \lambda_c + \sigma_{H_2,D_2} \tag{13}$$

$$\bar{v} = \tfrac{1}{6}\,\pi(\lambda_c + \sigma_{H_2,D_2})^3 \tag{14}$$

The calculated results are also given in Tab. 1.

The solubility coefficient, of gases in polymers, $S^{(i)}$, may be expressed as the mean number of gas molecules, $\bar{n}$, dissolved per cubic centimeter of polymer. This mean number, $\bar{n}$, times the mean volume per cavity, $\bar{v}$, is the total mean volume of H_2 or D_2 filled cavities per cubic centimeter of polymer $\Phi^{(i)}$. The solubility coefficients $S^{(i)}$ for H_2 and D_2 at 296 K and the corresponding total mean volumes Φ are given in Tab. 2.

Tab. 2. Solubilities of H_2 and D_2 in polymers at 296 K and the total mean volume of hydrogen isotope filled cavities

Polymer	Solubility S $\frac{\text{nr. gas molec.}}{\text{cm}^3\text{ polymer}} \times 10^{-17}$		Total mean volume Φ cm³ × 10^{-6}	
	H_2	D_2	H_2	D_2
PET	2·79	2·95	5·63	5·95
PE	3·57	4·14	7·67	8·90
PVC 1	7·52	7·93	15·9	16·7
PVC 2	6·72	7·12	17·4	18·4

According to an improved "Free Volume Theory" of gas transport through polymers the total fractional free volume v_{fs} (including voids) can be evaluated from diffusion and solubility coefficient measurements [10]. For example for PE this fractional free volume has a value of about $v_{fs} = 0{\cdot}025$, that is 0·025 cm³ of free volume per cubic centimeter of polymer [11]. Denoting by r the fraction of hydrogen isotope occupied free volume, i.e. $r = \frac{\Phi H_2}{v_{fs}}$ one finds, $r = 3{\cdot}5 \times 10^{-4}$ for PE. Although for the other studied polymers we have not found exact values of v_{fs}, a good approximation is to consider these fractional volumes, being in the range of 0·005–0·05 [12]. Thus, the corresponding r ratios will be in the range of $10^{-3}-10^{-4}$. The smallnes of this ratio allows one to understand why the diffusion coefficient of H_2 and D_2 through the studied polymers is concentration independent [13].

In order to obtain a rough indication of the number density of cavities in the polymer samples, ϱ_c, one must accept that the mean apparent hydrogen isotope occupied cavity volume is the same as the mean volume of a cavity in the polymer independently of whether if is occupied or not. For PE the number density of cavities $\varrho_c = \frac{v_{fs}}{\bar{v}}$ it $1{\cdot}2 \times 10^{21}$ cavities/cm³ polymer.

Hydrogen isotope transmission results through polymers

The existence of quantum isotope effects in the transport process of H_2 and D_2 in the studied polymers provides us with interesting informations concerning the mechanism of diffusion and the approximate geometrical dimension of the normal and of the transition states. At any given moment there exists within a polymer a spectrum of free volumes, or local density fluctuations, which are called cavities. They may appear as a result of the cooperative motion of several neighboring polymer macromolecules [14]. Obviously these cavities are not fixed in space, but on a time average and at a constant temperature they have more or less definite distribution in the spectrum of the free volumes. Only those cavities which are large enough to accomodate a penetrant molecule are effective in the gas transport through the polymer. It is also accepted that even if an individual cavity may not be large enough to accomodate a molecule, the cooperative motion of the polymer macromolecules may allow two or more cavities to merge into one large enough for a diffusional jump to occur. In our study we determined the mean hydrogen isotope accomodating cavity diameter d, Tab. 1. One can see that d is larger for PE than for PET respectively for PVC 2 than PVC 1. This situation could be correlated with the existing differences in the intramolecular cohesive forces in the polymers. In the first case the chain stiffness and cohesive energy of PE are smaller than those of PET, and consequently one expects that at a given temperature the thermal fluctuations of the polymer macromolecules generates more cavities in PE than in PET. The same mechanism may provide more cavities in PVC 2 than in PVC 1, but in this case the intramolecular cohesive forces and chain mobility in PVC are weakened by the increasing plasticizer content. As reported earlier [7], the differences in the number and size of the cavities and in the ease of cavity formation is also reflected in the absolute values of permeability, diffusion and solubility coefficients [8, 15]. These findings are in good agreement with the results from Tab. 1 and 2.

The diffusion coefficient $D^{(i)}$ of a gas through a polymer may be related to the jump frequency ν_i, of the molecule out of the cavities and the jump distance δ_i as follows [16]:

$$D^{(i)} = \tfrac{1}{6}\,\nu_i\delta_i^2 f \tag{15}$$

where: f is a correlation factor equal to the average value of the cosine of the angle between the first jump and the j-th following jump, $f = 1 + 2\sum_{1}^{n}\langle\cos\theta_j\rangle$. This factor has a value $f = 0{\cdot}781$ when the diffusion of molecules proceeds through vacancies [17]. In our study a rough estimate of the hydrogen isotope jump distance is $\delta_{H_2,D_2} = \lambda_b + \sigma_{H_2,D_2}$. The values obtained for the four polymers are summarized in Tab. 1. In the case of PE one can compare the obtained $\delta_{H_2,D_2} = 3{\cdot}65$ Å with the one calculated with a modified "Free Volume Theory" [18]. According to this theory the diffusion of H_2 or D_2 takes place along a somewhat cylindrical volume equal to:

$$V = \frac{\pi}{4}\sigma_{H_2,D_2}^2\lambda \tag{16}$$

where: λ is the length of the diffusional step. Assuming that the mean free volume V of H_2 in PE is about 50 Å³ [18] one obtains $\lambda = 4{\cdot}77$ Å. This value is larger than the one given in Tab. 1 for PE. As yet noticed [19] this difference may suggest that in the classical "Free Volume Theory" the critical volume necessary for a hydrogen molecule to move through the polymer is overestimated.

Taking into account that for PE, at 296 K, the mean jump distance is $\delta_{H_2,D_2} = 3{\cdot}65$ Å and the H_2 diffusion coefficient is $D_{H_2} = 58{\cdot}5 \times 10^{-7}$ cm²/s [8], the approximate jump frequency ν_{H_2} given by Eq. (15) has the value of about $\nu_{H_2} = 3 \times 10^{10}$ jumps per second. This value is considerably smaller than the frequency ν_v of the H_2 molecule vibration about its equilibrium position 10^{12} Hz $< \nu_v < 10^{13}$ Hz. Hence a H_2 molecule has many vibrations during a diffusional jump.

Quantum effects in H_2, D_2 and He diffusion through PET

Let us now focus our attention on the diffusion of some simple gases (including H_2, D_2 and He) through PET.

In the classical case, at a given temperature T, the diffusion coefficients D_T of a large number of gases through a polymer can be satisfactorily correlated by a relation of the form [20]:

$$\ln D_T = \ln D_{T_0} - \omega\sigma^n \tag{17}$$

where: D_{T_0} and ω are characteristic constants of the polymer, σ is the "effective molecular diameter" of the penetrant gas in the polymer, and n is a constant, which equals 1 for highly flexible chain macromolecules, such as natural rubber or PE, and 2 for rotationally hindered, stiff-chain macromolecules as PET or polycarbonate. Such a correlation for the diffusion coefficients D_T of Ne, Ar, N_2 and CO_2 through PET at 296 K has been reported in a previous paper [7] (Fig. 2).

With the experimental results plotted in Fig. 2 one can calculate from Eq. (17) the PET characteristic constants $D_{T_0} = 9{\cdot}3 \times 10^{-6}$ cm²/s and $\omega = -0{\cdot}517$ (Å²)⁻¹. Therefore, assuming that the He and H_2 effective diameters are $\sigma_{He} = 2{\cdot}556$ Å and $\sigma_{H_2,D_2} = 2{\cdot}928$ Å [9] one obtains from Eq. (17) the theoretical diffusion coefficients $D_{He}^{theor} = 3{\cdot}1 \times 10^{-7}$ cm²/s and $D_{H_2}^{theor} = 1{\cdot}08 \times 10^{-7}$ cm²/s, for He and H_2. These values are considerably smaller than the experimental ones: $D_{He}^{exp} = 12{\cdot}7 \times 10^{-7}$ cm²/s and respectively $D_{H_2}^{exp} = 5{\cdot}9 \times 10^{-7}$ cm²/s [8]. A similar situation was previously reported for polymers like; PE [21], natural rubber [22], polymethylmethacrilate [23] and polyvinyl acetate [24].

In order to explain these disagreements one must first analize the cavities involved in the H_2, D_2 and He diffusion through PET and then the influence of the energy quantization in these cavities on the H_2, D_2 and He diffusion coefficients.

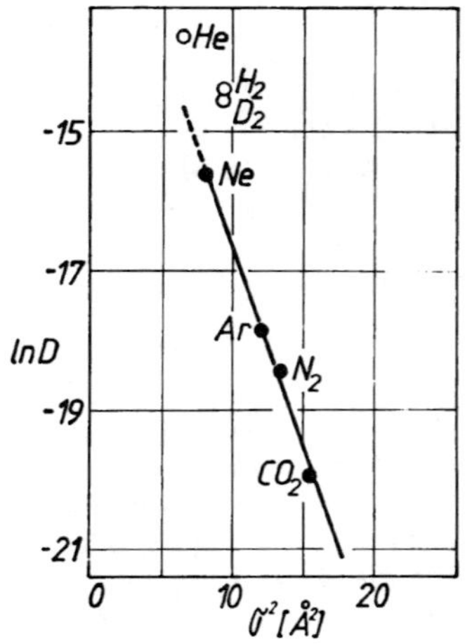

Fig. 2. Correlation of diffusion coefficients of gases in polyethylenetherephthalate at 296 K

As established the diffusion of H_2 and D_2 through PET takes place by means of jumps between the neighboring cavities k of mean diameter $d = 3{\cdot}38$ Å. Undoubtedly all these cavities may be involved also in the diffusion process of the smaller He. But He will also diffuse through the cavities q, large enough to accomodate He, but not H_2 or D_2. Thus, in the He diffusion process a larger number of cavities is involved and therefore the diffusion coefficient of He will be larger. In the classical case the ratio of these coefficients, as derived from Eq. (17), will be $D^{\text{theor}}_{\text{He}}/D^{\text{theor}}_{\text{H}_2} \cong 3$, larger than the experimentaly found one $D^{\text{exp}}_{\text{He}}/D^{\text{exp}}_{\text{H}_2} \cong 2$. Thus we must suppose that during the diffusion of He and H_2 through PET there exists a mechanism responsible for the increase of the diffusion coefficients and decrease of their ratio.

As stated earlier, for molecules such as H_2, D_2 and He, the quantization of translational energy inside the normal and activated states is no longer negligible. For H_2, D_2 and He the relative energy spacings in the same k cavity of PET are given in Fig. 3.

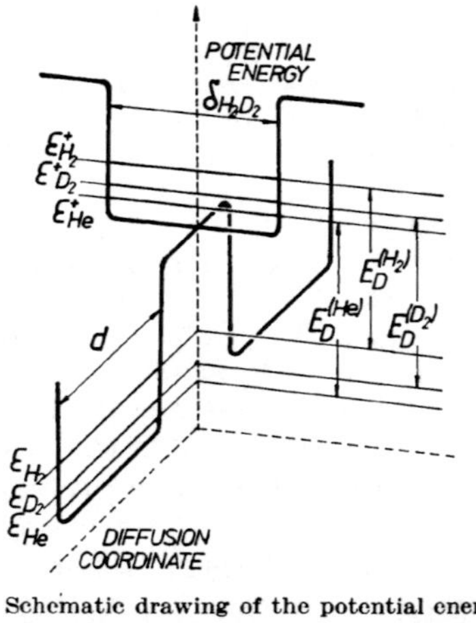

Fig. 3. Schematic drawing of the potential energy surface of k cavities in polyethylenetherephthalate and energy levels for H_2, D_2 and He

The translational energy of He in the Q_i and Q_i^+ states were calculated with Eq. (2) and (3) considering: $m_{\text{He}} = m_{\text{D}_2}$, $\lambda'_c = d - \sigma_{\text{He}} = 0{\cdot}86$ Å $> \lambda_c = 0{\cdot}45$ Å and $\lambda'_b = \delta_{\text{H}_2,\text{D}_2} - \sigma_{\text{He}} = 0{\cdot}96$ Å which is larger than $\lambda_b = 0{\cdot}59$ Å.

Therefore in Fig. 3:

$$\varepsilon_{\text{He}} = \frac{h^2N}{8m_{\text{He}}\lambda_c'^2} < \varepsilon_{\text{D}_2} = \frac{h^2N}{8m_{\text{D}_2}\lambda_c^2} < \varepsilon_{\text{H}_2} = \frac{h^2N}{8m_{\text{H}_2}\lambda_c^2} \tag{18}$$

$$\varepsilon^+_{\text{He}} = \frac{h^2N}{8m_{\text{He}}\lambda_b'^2} < \varepsilon^+_{\text{D}_2} = \frac{h^2N}{8m_{\text{D}_2}\lambda_b^2} < \varepsilon^+_{\text{H}_2} = \frac{h^2N}{8m_{\text{H}_2}\lambda_b^2} \tag{9}$$

In a thermaly activated diffusion process the jump frequency ν_i^k, of a molecule out of a cavity k with finite height decreases exponentialy with the increase of the energy gap to the top of the cavity [25]. In Eq. (15) the jump length σ_{He} between the k cavities is the same for He, H_2 and D_2. Therefore, the relative values of the diffusion coefficients will be determined by the relative values of the corresponding jump frequencies σ_i^k. From Eq. (18) one finds: $\nu^k_{\text{H}_2} > \nu^k_{\text{D}_2} > \nu^k_{\text{He}}$ which means that: $D^k_{\text{H}_2} > D^k_{\text{D}_2} > D^k_{\text{He}}$. In the case of H_2 and D_2 diffusion through PET corresponding to an energy level spacing of $\delta\varepsilon^+ - \delta\varepsilon = 0{\cdot}5$ kcal/mol one obtains a difference between the diffusion coefficients $\delta D^k = D^k_{\text{H}_2} - D^k_{\text{D}_2} = 1 \times 10^{-7}$ cm²/s [8]. Therefore the difference between the diffusion coefficients $\delta D'^k$ of H_2 and He through PET may be estimated from the energy level spacing of H_2 and He, $\delta\varepsilon' = 0{\cdot}8$ kcal/mol, accomodated in the same k cavity. The calculated value is $\delta D'^k = 1{\cdot}7 \times 10^{-7}$ cm²/s which leads to an improved theoretical He diffusion coefficient of about $D'^{\text{theor}}_{\text{He}} = 4{\cdot}3 \times 10^{-7}$ cm²/s. This value is approximatively one third of the experimental one, but it is larger than the value derived in the classical model $D^{\text{theor}}_{\text{He}} = 3{\cdot}1 \times 10^{-7}$ cm²/s. Nevertheless we must not forget that in deriving $D'^{\text{theor}}_{\text{He}}$ we considered that in the diffusion process of He are involved only cavities k. In order to reach the experimental value $D^{\text{exp}}_{\text{He}}$ we must admite that He diffuses also through cavities q. In these cavities the translational energy of He is also quantized and there will be a characteristic jump frequency ν^q_{He} of the He out the cavities q and a δ^q_{He} jump length between the neighboring cavities. Thus, according to Eq. (15) the total He diffusion coefficient through PET may be written as:

$$D^{\text{exp}}_{\text{He}} = \tfrac{1}{6} f[\nu^k_{\text{He}}(\delta^k_{\text{He}})^2 + \nu^q_{\text{He}}(\delta^q_{\text{He}})^2] \tag{20}$$

where: $\tfrac{1}{6} f[\nu^k_{\text{He}}(\delta^k_{\text{He}})^2] = D'^{\text{theor}}_{\text{He}} = 4{\cdot}3 \times 10^{-7}$ cm²/s and

$\tfrac{1}{6} f[\nu^q_{\text{He}}(\delta^q_{\text{He}})^2) = 8{\cdot}1 \times 10^{-7}$ cm²/s

Therefore from Eq. (24) one finds that, in the diffusion process of He through PET, the contribution of q cavities is twice larger than that of the k ones. Unfortunately with the available experimental data it is impossible to evaluate either the ν^q_{He} jump frequency or the δ^q_{He} jump length. Thus, at this stage we can not obtain valuable informations concerning the relative value of the density of k and q cavities in the polymer. If, $\nu^q_{\text{He}} < 2\nu^k_{\text{He}}$, $\delta^q_{\text{He}} > \delta^k_{\text{He}}$ and it may be supposed that the density of q cavities is smaller than of the k ones. On the contrary, if $\nu^q_{\text{He}} > 2\nu^k_{\text{He}}$, $\delta^q_{\text{He}} < \delta^k_{\text{He}}$ and the density of q cavities is larger than that of k cavities.

Conclusions

The presented quantum model for the transport of very light and small gas molecules through polymers, appears to offer a good explanation for the experimentaly reported isotope effects in the diffusion, permeation and solvation of H_2 and D_2 in four sorts of polymers. The measured isotope effects allowed us to obtain an approximate information about the geometrical dimensions of the normal and activated states of the polymers. The dependence of these dimensions on the characteristic properties of the polymers leads to results which are in good agreement with the model of simple gas diffusion through polymers [5, 13, 14].

Analyzing the diffusion of H_2, D_2 and He through a series of polymers it was found that their diffusion coefficients and activation energies of diffusion were larger respectively smaller than those derived by means of the classical theory of gas transport through polymers [7, 21 – 24].

As mentioned in the Introduction, according to this theory, whichever the diffusion molecule is, it performs a translational, vibrational and rotational movement on the bottom of the accomodating cavity [4, 26]. In the quantum picture the small and light H_2, D_2 and He molecules no longer move on the same energy level on the bottom of the cavity, but on higher discrete ones. Thus, the barrier height and the activation energy of diffusion for H_2, D_2 and He will be smaller and the corresponding jump frequencies and diffusion coefficients will be larger. As shown above this statement is in agreement with the results obtained in the study of H_2, D_2 and He diffusion through PET.

Considering the diffusion of larger and heavier simple gas molecules; N_2, O_2, CO_2, Ar, etc. through polymers, the quantization of translational energy in the normal and activated states is negligible. Therefore, one may expect that in the case of N_2 or O_2 isotopes transport through a polymer, one can not determine the isotope effects mentioned for H_2 and D_2.

Received March 14, 1985

References

[1] *J. Crank, G. S. Park*, Diffusion in Polymers. New York: Academic Press 1968

[2] *W. Jost*, Diffusion in Solids, Liquids, Gases. New York: Academic Press 1960

[3] *H. L. Frisch, C. E. Rogers*, J. Chem. Phys. **40** (1964) 2293

[4] *K. D. Ziegel, F. R. Eirich*, J. Polym. Sci., Polym. Phys. Ed. **12** (1974) 1127

[5] *V. T. Stannett, W. J. Koros, D. R. Paul, H. K. Lonsdale, R. W. Baker*, Adv. Polym. Sci. **32** (1979) 69

[6] *A. S. Nowick, J. J. Burton*, Diffusion in Solids. New York: Academic Press 1975

[7] *P. Mercea*, Rev. Chimie **33** (1982) 435

[8] *P. Mercea*, Isotopenpraxis **19** (1983) 153

[9] *J. O. Hirschfelder, F. G. Curtis, R. B. Bird*, Molecular Theory of Gases and Liquids. New York: Wiley 1954

[10] *F. Bueche*, J. Chem. Phys. **21** (1953) 1850

[11] *S. Matsuoka*, J. Appl. Phys. **32** (1961) 2334

[12] *P. Meares*, Trans. Faraday Soc. **54** (1958) 40

[13] *A. V. Tobolsky, H. Mark*, Polymer Science and Materials. New-York: Wiley 1971

[14] *H. L. Frisch*, Polym. Lett. **3** (1965) 13

[15] *P. Mercea*, Rev. Chimie **34** (1983) 806

[16] *J. R. Manning*, Diffusion Kinetics in Cristals. New-Jersey: Van Nostrand 1968

[17] *P. Meares*, J, Am. Chem. Soc. **76** (1954) 3415

[18] *C. A. Kumins, J. Roteman*, J. Polym. Sci. **55** (1961) 699

[19] *H. Eyring, D. Henderson, W. Jost*, Physical Chemistry, Vol. 10. New-York: Academic Press 1970

[20] *E. S. Perry*, Progress in Separation and Purification, Vol. 1. New-York: Interscience 1968

[21] *A. S. Michaels, H. J. Bixler*, J. Polym. Sci. **50** (1961) 413

[22] *G. J. Amerongen*, J. Polym. Sci. **5** (1950) 307

[23] *V. Stannett, J. L. Williams*, J-Polym. Sci., C **10** (1966) 45

[24] *P. Meares*, Trans. Faraday Soc. **53** (1957) 101

[25] *G. Vineyard*, J. Phys. Chem. Solids **3** (1957) 121

[26] *I. Prigogine*, The Molecular Theory of Solutions. North-Holland Amsterdam 1957

Isotopenpraxis **21** (1985) 12, S. 417 bis 423

Quantitative Analysis of Multicomponent Surface Layers by Means of XRF

V. Rößiger

(Zentralinstitut für Isotopen- und Strahlenforschung der AdW der DDR, Leipzig)[1])

X-ray fluorescence (XRF) analysis is used for the determination of thickness and composition of surface layers such as electroplated or evaporated coatings. The composition and thickness data evaluation from X-ray yields is done with the help of computer routines where the main prerequisite consists in an accurate calculation of the X-ray production for any given structur of the layer under consideration. Here the so-called enhancement effect plays an important role. In this paper a considerably simplified fundamental parameter treatment of secondary fluorecsence is derived and applied. It is demonstrated that XRF analysis even may be used for nondestructive depth profiling if the measured X-ray yields are influenced by the depth profile with sufficient sensitivity. In the case of strong interelement effects within the layer this technique may be applied successfully, as shown for FeNi foils as a typical example.

Die Röntgenfluoreszenzanalyse wird zur Bestimmung der Dicke und der Zusammensetzung von beispielsweise galvanisierten Legierungsschichten oder gedampften Überzügen eingesetzt. Die Grundlage dafür sind Rechenprogramme, denen die exakte Beschreibung der Röntgenstrahlungsproduktion für das jeweilige Schichtsystem als Ausgangspunkt dient. Die Interelementanregung (Enhancementeffekt) ist dabei von besonderer Bedeutung. Ihre Behandlung erfolgt hier in einer sehr einfachen und praktikablen Weise. Es wird außerdem gezeigt, daß sich die Röntgenfluoreszenzanalyse in bestimmten Fällen sogar zur Tiefenprofilmessung eignet. Als typisches Beispiel dazu diskutiert die Arbeit FeNi-Folien.

Keywords

electrochemical coating, layers, nondestructive analysis, spatial distribution, thickness gages, X-ray fluorescence analysis

Introduction

XRF analysis is widely used for the measurement of thickness and composition of surface coatings. Alloy coatings were analyzed [1 – 4] as well as sandwich-like solid films [5]. An important advantage of the XRF technique is, in principle, that it does not need any standards for calibration. This holds true if both X-ray production of a given sample (with defined thickness and composition)

[1]) Address: DDR-7050 Leipzig, Permoserstraße 15

and X-ray detection efficiencies are known with sufficient accuracy. The latter can be taken for granted if a well calibrated X-ray detector is used.

This paper deals with methodological aspects of the analysis of composed surface layers. The main feature in this context is the calculation of X-ray yields for given structure parameters of the sample, e.g. the thickness and concentration of elements in an alloy coating on a substrate. The interelement secondary X-ray excitation plays an important role here, and the different ways of treatment worked out for thick homogeneous samples [6—8] are not applicable for the surface coatings under consideration. In this paper a procedure of matrix correction is presented which is well-fitted to the specificity of the samples. It may be regarded as a rigorous approximation of the exact fundamental parameter treatment for specimens of intermediate thickness [1, 9] without any substantial loss of accuracy.

The functional dependence of the X-ray yields on the structure parameters is relatively complicated due to the complexity of excitation and absorption processes which have to be taken into account. Hence, a reversion of these nonlinear relations is not possible in general, and numerical methods have to be applied in order to determine concrete structure parameter values from experimental data.

1. Calculation of XRF yields

To simplify matters, monoenergetic excitation is assumed in this paper. In many cases it is possible to realize this by experimental means such as a suitable X-ray source combined with absorption filters or secondary fluorescent targets. Otherwise the polychromatic or continuous primary distribution may be treated as a superposition of discrete energy components. In order to describe the principles of X-ray fluorescence of any structure of surface layers we outline the X-ray yield calculation of a flat binary alloy film. Other imaginable cases of practical importance may be treated in a similar way without having to take into account additional physical phenomena.

1.1. *Excitation by primary X-rays*

The X-ray fluorescence yield of a fluorescent component I concerning to the element i is given by

$$Y_{\mathrm{I}} = \frac{C_{\mathrm{i}} \cdot \tau_0^{\mathrm{I}}}{\cos\varphi} \cdot \frac{1 - \exp(-\mu' X)}{\mu'} \tag{1}$$

with

$$\mu' = \frac{\mu_0}{\cos\varphi} + \frac{\mu_{\mathrm{I}}}{\cos\psi} \tag{2}$$

Here the X-ray yield Y is defined per one incident primary photon of the energy E_0. Furthermore, the detection efficiency is omitted in order to simplify matters. Therefore, the measured yield is given by $n \cdot \varepsilon \cdot (\Omega/4\pi) \cdot Y$, with $n =$ number of incident primary photons, $\varepsilon =$ relative efficiency of the detector, and $\Omega =$ its solid angle. The binary alloy layer (elements i and j) is characterized by the concentrations $C_{\mathrm{i}} = 1 - C_{\mathrm{j}}$ and its thickness X. The attenuation cross sections μ are labelled by an index for the energy ($0 \mathrel{\hat{=}} E_0$; $I \mathrel{\hat{=}}$ energy of the fluorescent component under consideration). In order to avoid the density ϱ in the formulas the dimension of X is at/cm² or g/cm² which corresponds to at/at or g/g for the concentration and to cm²/at or cm²/g for the cross sections, respectively. τ_0^{I} denotes the cross section for the production of an event of the kind I for the primary energy. It has to be calculated from the ionization cross section τ_0 according to

$$\tau_0^{\mathrm{I}} = P(I) \cdot \omega(I) \cdot w(I) \cdot \tau_0 \tag{3}$$

$P(I) = P_{\mathrm{K}}, P_{\mathrm{L}} \ldots$ denotes the probability of ionizing the respective atomic shell K, L, ..., and $\omega(I) = \omega_{\mathrm{K}}, \bar{\omega}_{\mathrm{L}}, \ldots$ means the fluorescence yield of this shell. Furthermore, $w(I) = w_{\mathrm{K}_\alpha}, w_{\mathrm{K}_\beta} \ldots$ is the probability of emitting a certain component of characteristic radiation within the K-, L-, ... series.

φ and ψ are the angle of primary beam incidence and the take-off angle of the fluorescent radiation with respect to the sample normal (cf. Fig. 1).

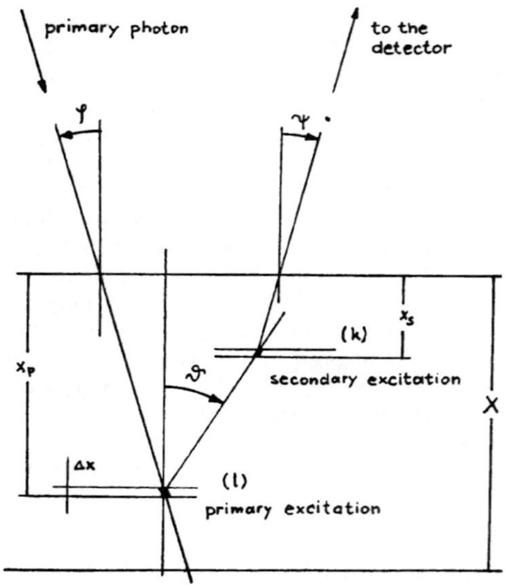

Fig. 1. For the derivation of secondary fluorescence yield Y_{I}'

In addition to this, it has to be mentioned that the attenuation cross sections are weighted averages according to

$$\mu = C_{\mathrm{i}} \cdot \mu^{\mathrm{i}} + C_{\mathrm{j}} \cdot \mu^{\mathrm{j}} \tag{4}$$

where the upper indices characterize the element. Finally, for the special cases of a thick and a thin layer ($X \to \infty$, $X \to 0$) Eq. (1) leads to the well-konwn equations

$$Y_{\mathrm{I}}^{\mathrm{thick}} = \frac{C_{\mathrm{i}} \cdot \tau_0^{\mathrm{I}}}{\cos\varphi \cdot \mu'} \tag{5}$$

and

$$Y_{\mathrm{I}}^{\mathrm{thin}} = \frac{C_{\mathrm{i}} \cdot \tau_0^{\mathrm{I}} \cdot X}{\cos\varphi} \tag{6}$$

respectively.

1.2. *The enhancement effect*

The X-ray excitation of the fluorescent component I (e.g. the K_α of element i) by characteristic radiation J (e.g. K_α of element j) causes a term Y_{I}' in addition to the yield Y_{I} of Eq. (1). Its order of magnitude with respect to Y_{I} depends on the sample composition and the excitation mode. Y_{I}' vanishes if there is no fluorescent component of the element j with an energy high enough to ionize the atomic shell relating to the component I. Y_{I}' becomes maximum for an exciting component J with an energy near the respective excitation threshould (absorption edge).

The general situation for calculating the secondary fluorescence yield Y_{I}' is sketched in Fig. 1. A primary photon acts on the sublayer l at the depth x_p exciting a certain fluorescence yield of the J-type fluorescence radiation $\Delta y_J(l)$. This secondary X-ray source in turn excites the fluorescence radiation of I-type in the sublayer k having the distance $|x_p - x_s|$ from l. The respective yield is denoted with $\Delta y_I'(k, l)$. For its exact calculation one has to take into account all azimutal angles ϑ running from 0° to 90° for $x_s \leq x_p$, and from 90° to 180° for $x_p < x_s < X$ (= total thickness of the layer), respectively. For obtaining the total secondary yield Y_I' one has to carry out a double-summation of $\Delta y_I'(k, l)$ with respect to all sublayers k and l. If Δx becomes infinitesimal ($\Delta x \rightarrow \mathrm{d}x$) the sum is converted into an integral over x_s and x_p. The latter is calculated in [1] where the ϑ-integration cannot be made by elementary means.

The first approximation of this work consists in a subdividing the layer into a number of L sublayers of constant thickness $\Delta x = X/L$ where Δx is much less than the relaxation length of the enhancing radiation ($L \gg X \cdot \mu_J$). In every layer l ($l = 1 \ldots L$) the sources of the enhancing radiation of the J-type are localized as point sources in its centre. Their intensities $\Delta y_J(l)$ have to be calculated according to Eq. (1)

$$\Delta y_J(l) = \exp\left(-\frac{(l-\frac{1}{2}) \cdot (X/L) \cdot \mu_0}{\cos\varphi}\right) \cdot Y_J(X/L) \tag{7}$$

where the exponential term accounts for primaries absorption from the surface to the l-th sublayer (cf. Fig. 1). The second approximation is of a more drastic character. The isotope X-ray emission from the l-th sublayer (which causes the integration over ϑ in order to calculate $\Delta y_I'(k, l)$) is replaced by an emission under two fixed angles ϑ_1 and ϑ_2 into one half-space ($\vartheta = 0° \ldots 90°$) with the relative intensities a and $(1 - a)$, respectively. Section 1.3. justifies this approximation in detail and derives the values of the parameters:

$$\vartheta_1 = 47°\,, \qquad \vartheta_2 = 82° \quad \text{and} \quad a = 0{\cdot}66$$

Now the action of the J-component of every sublayer l on the k-th sublayer is simply described by two angular terms. The yield of secondary fluorescence is given by

$$\Delta y_I'(k, l) = \frac{C_i \cdot \tau_J^I \cdot X}{2L} \cdot \Delta y_J(l) \cdot \left\{\frac{0{\cdot}66 \cdot A(k, l, \vartheta_1)}{|\cos \vartheta_1|} + \frac{0{\cdot}34 \cdot A(k, l, \vartheta_2)}{|\cos \vartheta_2|}\right\} \tag{8}$$

The factor A describes the absorption of the J-type X-rays from k to l

$$A(k, l, \vartheta) = \exp\left(-\frac{|k - l| \cdot (X/L)}{|\cos \vartheta|} \cdot \mu_J\right) \tag{9}$$

The total enhancement contribution is given by a superposition of all $\Delta y_I'(k, l)$ ($k = 1 \ldots L$) taking into account absorption from the k-th sublayer to the surface:

$$Y_I' = \sum_{k=1}^{L} \sum_{l=1}^{L} \exp\left(-\frac{(l-\frac{1}{2})\,(X/L)}{\cos\psi} \cdot y_I\right) \Delta y_I'(k, l) \tag{10}$$

It has to be added to the yield from primary excitation (Eq. (1)) where, of course, all enhancing components J have to be taken into consideration.

These two simplifications are used in practical XRF analysis of surface layers and foils as described in Section 2.

The first one (summing-up instead of integration) is of importance if a depth-dependent concentration function is taken into account as discussed in Section 2.4. For homogeneous samples (substrates) a sufficiently high L or, finally, integration may prevent any coarsing effect.

1.3. The approximation of isotope X-ray emission by discrete angular components

This section derives a justification for it. In order to find an accurate two angle approximation of the isotrope X-ray emission of a point source S into a half-space we look at the number of photons j which pentrate a plane P being a distance x away from the point source (cf. Fig. 2). It is given by

$$j(x) = (j_0/2) \cdot \int_0^{90°} \exp\left(-\mu x/\cos\vartheta\right) \cdot \sin\vartheta\, \mathrm{d}\vartheta \tag{11}$$

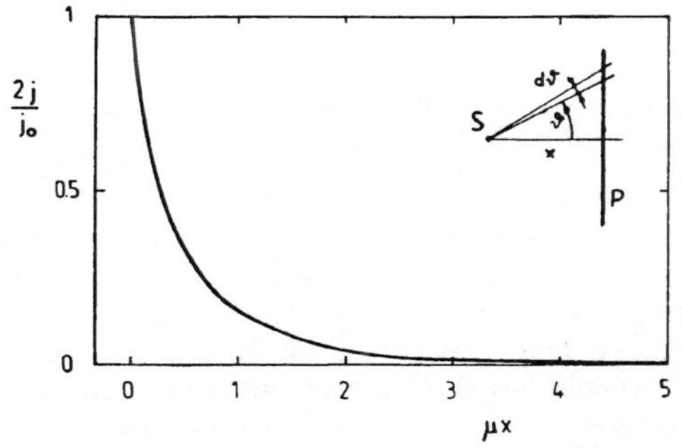

Fig. 2. X-ray flux $j/(j_0/2)$ as a function of μx according to Eq. (13)

where μ denotes the attenuation cross section of the medium. For vanishing attenuation ($\mu = 0$) j is constant and equal to one half of the total emission j_0 of the source. On the other hand, if the photons are emitted under two fixed angles ϑ_1 and ϑ_2 into a half-space we find

$$j'(x) = (j_0/2) \cdot \{a \cdot \exp(-\mu x/\cos\vartheta_1) + (1 - a) \cdot \exp(-\mu x/\cos\vartheta_2)\} \tag{12}$$

Here a, or $(1 - a)$ is the relative weight of the component with ϑ_1 or ϑ_2, respectively. If one takes into account one angle component only a becomes unity and the second term in Eq. (12) vanishes. The parameters ϑ_1, ϑ_2 and a of $j'(x)$ should be determined in such a way that $j'(x)$ is in best agreement with the exact function $j(x)$ according to Eq. (11). Following this concept, at first one has to calculate the integral of Eq. (11) by the well-known procedure. Substituting $u = -\mu x/\cos\vartheta$ we find

$$j(x) = (j_0/2) \cdot \mu x \int_{-\infty}^{-\mu x} \frac{\exp(u)}{u^2}\, \mathrm{d}u$$

and

$$j(x) = (j_0/2) \cdot \{\exp(-\mu x) + \mu x \cdot \mathrm{Ei}(-\mu x)\} \tag{13}$$

The exponential integral $\mathrm{Ei}(-\mu x)$ is numerically calculated by the progression

$$\mathrm{Ei}(y) = C + \log|y| + \sum_{i=1}^{\infty} \frac{y^i}{i \cdot i} \tag{14}$$

with *Eulers* constant $C = 0{\cdot}5772\ldots$.

The result is sketched in Fig. 2. This function $j(x)$ was taken to evaluate the parameters ϑ_1, ϑ_2 and a (cf. Eq. (12)) by a least square technique applying the SIMPLEX method [10]. The model function $j'(x)$ was fitted at $\mu x = 0{\cdot}1$, $0{\cdot}2, \ldots, 5{\cdot}0$. The best agreement was obtained for $\vartheta_1 = 47°$ with a relative weight $a = 0{\cdot}66$ and $\vartheta_2 = 82°$ and a relative

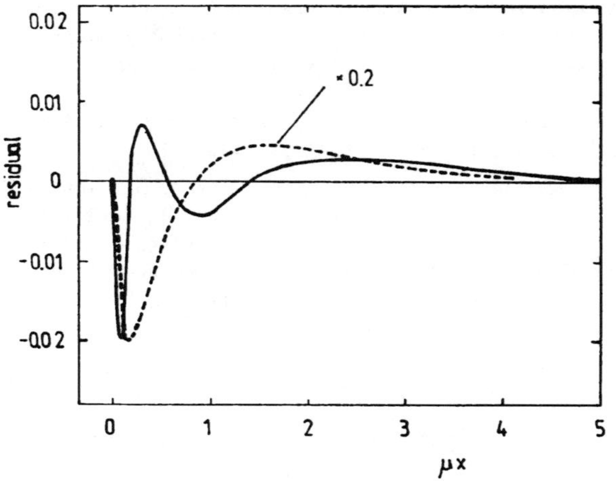

Fig. 3. Residuals $(j(x) - j'(x))/(j_0/2)$ for fitting $j(x)$ (Eq. (13)) by the approximation $j'(x)$ (Eq. (12)) with $a = 0{\cdot}66$, $\vartheta_1 = 47°$ and $\vartheta_2 = 82°$ (straight line) and with $a = 1$, $\vartheta_1 = 60°$ (dashed line)

weight 0·34, respectively. The residual $(j(x) - j'(x))/(j_0/2)$ is given in Fig. 3. It indicates a maximum error of fitting the exact flux of about 2 per cent at $\mu x \approx 0{\cdot}15$. The effect of those deviations on X-ray production calculation depends on the concrete situation, and it is expected to have the same order of magnitude. In this context it should be mentioned that secondary excitation becomes effective for $\mu x \gtrsim 1$, and in this thickness region the differences partly cancel each other.

In addition to the two angle approximation, Fig. 3 shows the residual of a one angle approximation ($a = 1$) with $\vartheta_1 = 60°$. It is valuable not only estimating the enhancement effect but also for its numerical description as such if its share in total X-ray production is small (less than about 5 ... 10 per cent, for orientation).

2. Applications

In this section selected typical examples of practical application are discussed which partly serve as experimental verification of the suggested matrix correction formalism. Due to the numerically effective handling of the interelement excitation it is possible to introduce polychromatic excitation (integration with respect to the primary energy E_0) and/or cylindrical samples (integration with respect to the angles of incidence φ and take-off ψ) without taking up too much computing time. Another specificity is depth profiling (Section 2.4.) for which this treatment of secondary fluorescence is well-fitted.

2.1. *Thickness gauging of pure element coatings*

This relatively simple case becomes interesting if a strong enhancement of a fluorescent component of the layer takes place due to substrate fluorescence radiation. It is discussed now for thin Cr coatings on Cu substrate in order to give a first example of matrix correction with sufficient accuracy. Fig. 4 shows the CrK_α yield as a function of Cr thickness with and without taking into account secondary excitation. The contributation of the enhancement effect is very large here: more than 100 per cent for small Cr thickness. It was calculated by the two angle approximation of Section 1: the CuK_α and CuK_β thick target yields for ϑ_1 and ϑ_2 as take-off angles act on the Cr layer. Tab. 1 gives the results obtained by different XRF modifications:

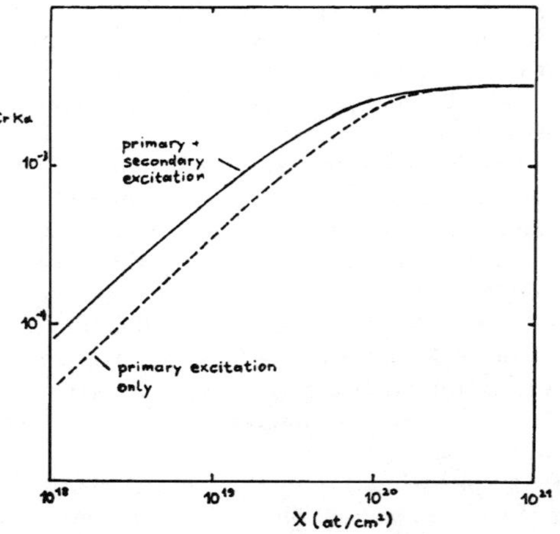

Fig. 4. CrKα X-ray yield of a Cr coating on Cu substrate as a function of layer thickness X calculated according to Section 1 (excitation with AgK, $\varphi = \psi = 45°$)

This curve was taken for thickness evaluation of the specimens of Tab. 1, column 4 (for conversion of thickness units: 10^{19} at/cm² ≙ 9·0 10^{-4} g/cm² ≙ 1·25 μm).

Tab. 1. Results of thickness gauging of Cr coatings on Cu for X-ray absorption (colums 1 and 2) and X-ray fluorescence (columns 3 – 5) analysis

primary radiation source:	^{238}Pu	^{238}Pu	^{238}Pu	Ag-tube	^{55}Fe
type of primary radiation	UL	UL	UL	AgK	MnK
	1	2	3	4	5
No.	Cr thickness Xρ, μm				
1	1·94	1·90	1·97	1·97	2·19
2	0·36	0·39	0·36	0·36	0·36
3	0·33	0·31	0·27	0·24	0·26
4	1·10	1·12	0·99	1·00	0·95
5	0·98	1·02	0·93	0·89	0·86
6	0·23	0·36	0·27	0·25	0·25

excitation by AgK (Ag target tube, Ag filter), ^{238}Pu (radioactive source which emits UL X-rays), and ^{55}Fe (radioactive source which emits MnK X-rays). The latter one excites Cr only by MnK_β (no enhancement possible). The respective thickness-yield-relation was taken from a previous calibration procedure [5]. The first two columns refer to the CuK attenuation due to the Cr layer and are completely independent of secondary excitation.

In column 1 the CuK_α yield with respect to the CuK_α yield for the absent Cr layer is utilized and column 2 interprets the ratio $Y_{CuK_\alpha}/Y_{CuK_\beta}$ in terms of Cr thickness.

The close agreement of the results of different methods and especially within the XRF modifications represents a good verification of the matrix correction including the suggested two angle approximation.

2.2. *Determination of thickness and composition of FeNi foils*

Electrodeposited FeNi foils (coatings were removed from the substrates) are to be analyzed for thickness and mean composition and, in addition, the composition depth profile for a homogeneity test. The latter topic is discussed in Section 2.4. Here we deal with XRF analysis of FeNi foils assuming constant composition with respect to the depth coordinate (which was confirmed to a certain degree by the above mentioned profiling investigations). For AgK excitation the FeK_α and NiK_α yields were interpreted in terms of composition and thickness applying a computer program based on the principles of Section 1. In addition, the alloy foils were analyzed by non-XRF methods and the results obtained may be used for a further comparison and verification. Both thickness and mean composition may be determined by an absorption technique [11]. It utilizes the selective effect of the absorption edges of Fe and Ni with respect to CuK_α. In the present case a special secondary fluorescence target (^{241}Am for primary excitation) of Cu and Mo was used, and the results were directly obtained from the attenuation of CuK_α and MoK_α caused by the foil to be analyzed. The sensitivity of this method is lower than that of XRF analysis whereas the data interpretation is much simpler. Especially, there is no influence of the elements depth distribution which is important for the analysis of nonhomogenious solid films. Tab. 2 and 3 give a compliation of composition analysis data for different methods, with Table 2 showing only the agreement of the absorption technique with conventional methods. Tab. 4 compares thickness results. The experimental errors occuring for several reasons may explain the differences of some per cents which are consistent with the demands of application in the present case. The general good agreement demonstates the reliability of the methods used for analysis, and especially the present version of the XRF technique is confirmed.

Tab. 2. Results of composition analysis of FeNi foils for different methods: comparison of the absorption technique with conventional chemical methods of analysis

method of analysis	AAS	complexometry	X-ray absorption [11]
No.	1	2	3
	Fe concentration, 10^{-2} g/g		
1	22·8	20·9	20·5
2	23·3	20·5	20·4
3	31·4	29·3	28·2
4	29·4	26·2	26·3
5	25·2	22·6	23·1
6	22·8	21·2	23·2
7	21·8	20·9	20·5

Tab. 3. Results of composition analysis of FeNi foils for different methods: comparison of the absorption technique (column 1) with XRF analysis (column 2) and with XRF depth profiling (column 3 gives averaged Fe concentrations, cf. Section 2.4.)

No.	1	2	3
	Fe concentration, 10^{-2} g/g		
1(≙A)	61·0	—	59·5
2(≙B)	18·9	—	20·5
3	21·7	22·2	22·9
4	20·1	20·8	21·1
5	21·0	22·3	22·0
6	19·2	23·2	22·1
7	25·9	24·2	24·2
8	24·8	25·4	23·9

Tab. 4. Results of thickness analysis of FeNi foils for different methods: comparison of the absorption technique (column 1) with XRF (column 2) and gravimetrical (column 3) thickness determination

No.	1	2	3
	thickness, 10^{-3} g/cm^2		
1	6·4	6·3	6·4
2	7·2	7·4	7·6
3	8·0	7·9	8·3
4	7·6	6·6	7·2
5	6·7	6·2	6·4
6	6·0	5·7	6·4
7	7·3	7·4	7·7

2.3. *Determination of thickness and composition of alloy coatings on a substrate*

Usually the homogeneity of the layer may be assumed and XRF analysis is utilized for determination of thickness and composition. In certain cases we have extented these investigations by PIXE experiments [12, 13]. This method is a valuable completion of XRF analysis. It may be used for surface near composition analysis (profiling) and/or for detection of surface contaminations.

As an example, we have investigated steel wire coated with a thin brass layer used for cord wire in tyres. Thickness and mean composition were measured by means of XRF analysis [4] and the surface near CuZn depth profile was studied utilizing the PIXE technique [13].

The most simple case in analyzing alloy coatings is characterized by monoenergetic excitation and plane samples. Polychromatic excitation and curved (e.g. cylindrical) specimens do not change the pricipal handling of the problem. As an illustrative example NiZn coatings on Fe shall be discussed. Fig. 5 depicts the X-ray dependence of the K_α yields both for substrate and coating elements. It can be seen that the ratios $Y_{NiK_\alpha}/Y_{FeK_\alpha}$ and $Y_{ZnK_\alpha}/Y_{FeK_\alpha}$ are very sensitive with respect to thickness and composition (see also Fig. 6). These yield ratios may be measured without difficulty, and they are a suitable basis for a rapid alloy coating analysis. The use of intensity ratios instead of absolute yields is favourable because there is no need for a calibrated primary X-ray flux with long term stability. Otherwise it would be possible to evaluate the unknown parameters from absolute yields Y_{NiK_α} and Y_{ZnK_α} alone. Unfortunately, in both cases the relationships between the

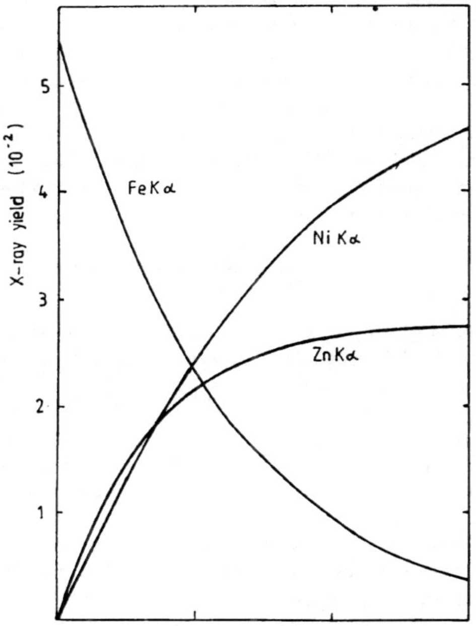

Fig. 5. K_α X-ray yields of a NiNn coating on Fe as a function of layer thickness X for $C_{Ni} = C_{Zn} = 0{\cdot}5$ (excitation with AgK, $\varphi = \psi = 45°$)

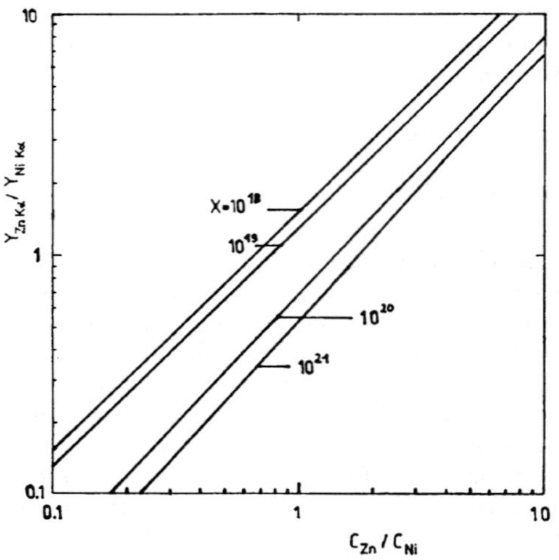

Fig. 6. X-ray intensity ratio $Y_{ZnK\alpha}/Y_{NiK\alpha}$ as a function of the concentration ratio C_{Zn}/C_{Ni} for different values of layer thickness X

experimental data on the one hand and the quantities to be determined (e.g. X and C_{Zn}) on the other is extremely nonlinear. For example, in the simple relation

$$\frac{C_{Zn}}{C_{Ni}} = d(X) \cdot \frac{Y_{ZnK_\alpha}}{Y_{NiK_\alpha}} \tag{15}$$

the factor d strongly depends on X as shown in Fig. 6. Hence, the data evaluation has to be performed with the help of a computer programm including an iterative strategy (similar to the LAMA program of [1]) which solves the system of nonlinear equations

$$\left.\begin{aligned} X &= u(Y_{NiK_\alpha}, Y_{ZnK_\alpha}) \\ C_{Zn} &= v(Y_{NiK_\alpha}, Y_{ZnK_\alpha}) \end{aligned}\right\} \tag{16}$$

Linear approximations as used in [3, 4] (for thin CuZn layers) are valid in a small thickness range only. Exponential series representations or other analytical approximations of the functions u and v are then necessary to reach a sufficient accuracy. Wether such model functions should be used for straightforward calculation, instead of the above mentioned iterative procedures, depends on the computers available. The loss of accuracy due to the considered approximations has to be weighted against the rapid computation of the result.

2.4. *Evaluation of concentration depth profiles of FeNi foils*

Now an attempt is made at depth profiling as a parameter fit procedure. We consider the depth-dependent concentration function of a binary alloy (i, j) $C_i(x) = 1 - C_j(x)$, with $x = 0 \dots X$. It is approximated with the help of a suitable model function $C_i'(p_1 \dots p_N; x)$ containing the free parameters $p_1 \dots p_N$. The task consists in the determination of these parameters from experimental X-ray yields with sufficient accuracy. The evaluation of them is to take place similar to the simple two parameter problem of thickness and composition determination (cf. Section 2.3. where the number of equations in the system (16) has to be enlarged considerably). Besides the questions of numerical parameter fit, the main problem consists in the calculation of the X-ray yields under consideration for any fictive profile function $C_i'(p_1 \dots p_N; x)$. Only in a next step the parameters $p_1 \dots p_N$ have to be shifted in such a way that the experimental and calculated X-ray yields or yield ratios reach the best agreement. That set of parameters which gives the best agreement is regarded as the solution to the problem.

In order to calculate the X-ray yield of a fluorescent component I emitted by the element i one has to start with the yield equation

$$\begin{aligned} Y_I(E_0, \varphi, \psi) &= \int_0^X [\exp(-\mu_0(x) \cdot x/\cos\varphi)] \\ &\times [\tau_0^I/\cos\varphi + S(x)] \cdot [\exp(-\mu_I(x) \cdot x/\cos\psi)] \\ &\times C_i(x)\, dx \end{aligned} \tag{17}$$

The term S takes into account the interelement excitation. In order to apply the principles of matrix correction of Section 1 the integral has to be replaced by a sum which is necessary for carrying out the yield calculation:

$$\begin{aligned} Y_I(E_0, \varphi, \psi) &= \sum_{l=1}^{L} \{(\Delta y_I(l) + \sum_{k=1}^{L} \Delta y_I'(k, l)) \\ &\times \exp\left(-\frac{(l - \frac{1}{2}) \cdot (X/L)}{\cos\psi} \cdot \mu_I(x)\right) \end{aligned} \tag{18}$$

In contrast to Section 1, the attenuation cross sections are functions of x propagated by the profile $C_i(x)$, cf. Eq. (4). Both the absorption of primary radiation (μ_0) and of fluorescent radiation (μ_I, μ_J) take place at different places with different sample compositions. Introducing the sum representation of Eq. (18) one has to take an approximation $C_i'(p_1 \dots p_N; x)$ of the depth profile which is well-fitted to the subdividing the layer into L sublayer with constant concentration. For example, this may be done by applying a step function with N equidistant steps; and L is 2, 3, or 4 times N.

For FeNi foils (thickness about 10 µm) the yield ratio

$$g = \frac{Y_{FeK_\alpha}}{Y_{NiK_\alpha}} \tag{19}$$

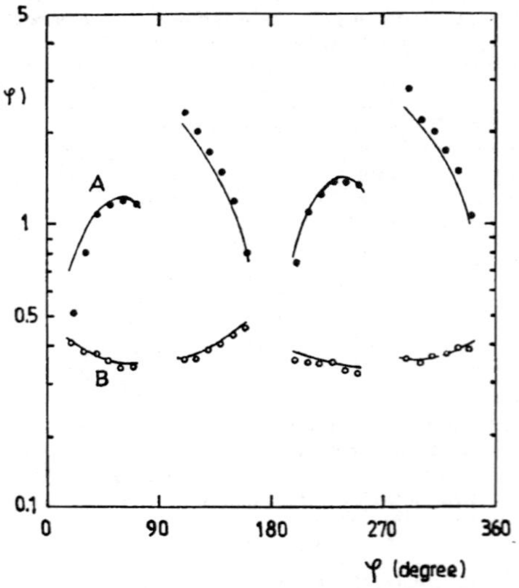

Fig. 7. X-ray yield ratio g as a function of the entrance angle φ ($\psi = \varphi - 90°$) for two samples A (closed circles) and B (open circles)

Sample A is a sandwich like Ni-Fe-Ni structure made by electroplating an Fe foil with Ni on both sides. Sample B is an electroplated alloy coating. It was removed from the substrate and analyzed as a foil in order to enable the angle variation over 360 degrees. The curves in the diagram correspond to the profiling results (cf. Fig. 8 and 9) (excitation with AgK).

was measured as a function of entrance angle φ ($\varphi = 20° \ldots 340°$, cf. Fig. 7). The free parameters for profile evaluation were the single Fe concentrations of a step function with N equidistant steps. The thickness X of the samples was determined by the absorption technique [11] mentioned already in Section 2.2., independently of the X-ray emission measurement. This method also yields the averaged Fe content of the foil which is a valuable test of the profiling results (cf. Tab. 3).

For a given (fictive) set of parameters $p_1 \ldots p_N$ the yields of Eq. (19) were calculated according to Eq. (18) applying the treatment of the enhancement effect outlined in Section 1. This has to be done for every measurement characterized by the angle φ_m, $m = 1 \ldots M$ (here $M = 24$). If one denotes the calculated yield ratio by g'_m and the experimental one by g_m^{exp} the solution to the problem is given by the minimum of

$$F = \sum_{m=1}^{M} \left(\frac{g'_m - g_m^{exp}}{g'_m} \right)^2 \qquad (20)$$

The minimum of F with respect to the parameters $p_1 \ldots p_N$ may be found by utilizing the SIMPLEX method [10]. Fig. 7 shows the experimental and the best fit yield ratios g_m of two typical samples, and in Fig. 8 and 9 the corresponding profile approximations are outlined.

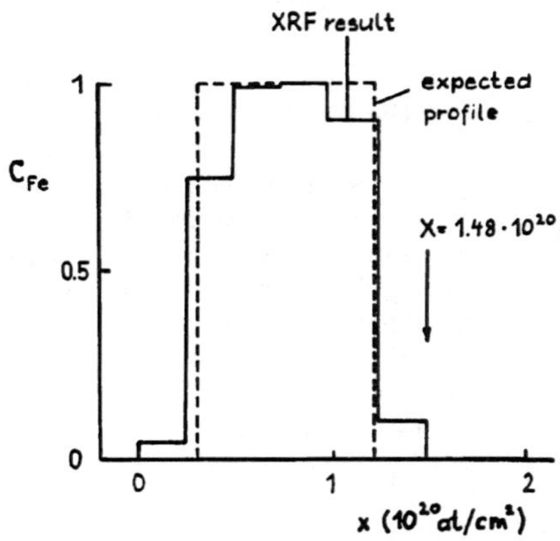

Fig. 8. Best fit profile result on the basis of the experimental yield ratios (cf. Fig. 7) of sample A (Ni-Fe-Ni sandwich)

As a model function for the Fe profile a step function with 6 steps of equal thickness was applied. The dashed line depicts the expected shape of the profile: sandwich structure known from sample proparation, $\bar{C}_{Fe}$ from the absorption measurement (cf. Section 2.2., Tab. 3), the small centroid shift corresponds to the centroid shift of the XRF profiling result.

The numerical fitting procedure was started with a constant profile. It was carried out by the computer without any knowledge of the expected shape of the depth profile. Therefore, the result of sample A (Ni-Fe-Ni sandwich) is a good experimental proof of the XRF profiling technique itself. Sample B (Fig. 9) is typical for the application of the method where alloy plating was to be controlled for constant Fe content at any depth. To provide an additional proof, PIXE measurements were carried out for the determination of surface composition at different depth (material removal by Ar sputtering).

Finally, the general problems of the XRF technique are quite similar to nondestructive PIXE or XPS depth profiling [13—15]. The questions connected with the choice of the model function $C'(p_1 \ldots p_N; x)$, its number of free parameters, the reliability and accuracy of its determination are discussed in [13, 14] dealing with PIXE. In this context the X-ray excitation by charged particles (PIXE) and by X-rays (XRF) act in a similar manner. In all cases, the depth distribution of a concentration is reflected by a series of experimental data. Of course, this reflection is of an extremely implicit character, as can be seen from a comparison of Fig. 7 (experimental data) with Fig. 8 and 9 (evaluated profiles). Due to the limited precision of measurements and the limited accuracy of the physical model the depth information is relatively poor in contrast to those profiling techniques working with a removal of material by etching. Nevertheless, there are some advantageous applications of the XRF technique.

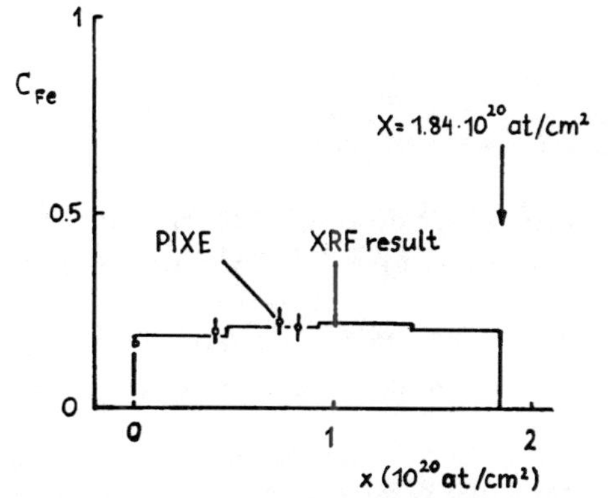

Fig. 9. Fe depth profile of an electroplated FeNi foil (sample B) evaluated on the basis of the experimental yield ratios of Fig. 7 applying a step function with 4 steps

The PIXE data also shown in the diagram were obtained from measurements with low energy protons (300 keV) in order to get an information depth much less than the thickness of the layer under consideration (cf. [13]).

Summarizing, Fig. 7—9 give a representative impression of how the method works. The general behaviour of the profile function is reflected by the XRF profiling results with good reliability, whereas details are not "seen" due to the poor depth resolution as is shown by the comparison with the expected (true) profile in Fig. 8.

Acknowledgement

The author is indepted to Doctor *R. Tertian* for helpful critical remarks and to Professor *G. Brunner* for constructive discussions. In addition, he and Mrs. *U. Schroth* have contributed some experimental data (Tab. 2 and 3) which is thankful acknowledged.

Received April 16, 1985

References

[1] *D. Laguitton, W. Parrish*, Anal. Chem. **49** (1977) 1152
[2] *R. Shan, A. Sandberg*, Advances in X-Ray Analysis, Vol. 26, p. 431, New York: Plenum 1983
[3] *J. L. Ponchon, P. Bourrain, A. Palsky*, X-Ray Spectrom. **12** (1983) 79
[4] *D. Gollan, G. Haschke, V. Rößiger*, Metalloberfläche **37** (1983) 496
[5] *V. Rößiger*, Isotopenpraxis **18** (1982) 338
[6] *N. Broll, R. Tertian*, X-Ray Spectrom. **12** (1983) 30
[7] *J. W. Criss, L. S. Birks, J. V. Gilfrich*, Anal. Chem. **50** (1978) 33
[8] *B. Van Oystaeyen, G. Demortier*, Nucl. Instr. Meth. **215** (1983) 299
[9] *D. Laguitton, M. Mantler*, Advances in X-Ray Analysis, Vol. 20, p. 515. New York: Plenum 1977
[10] *J. A. Nelder, R. Mead*, Computer J. **7** (1965) 3
[11] *V. Rößiger*, Metalloberfläche **36** (1982) 380
[12] *G. Brunner, V. Rößiger*, Nucl. Instr. Meth. B **3** (1984) 275
[13] *V. Rößiger, M. Lenk*, X-Ray Spectrom. **14** (1985) 62.
[14] *V. Rößiger*, Nucl. Instr. Meth. **196** (1982) 483
[15] *M. Pijolat, G. Hollinger*, Surface Sci. **105** (1981) 114

Isotopenpraxis 21 (1985) 12, S. 424 bis 429

Simplified Theory of the Mixing Effects of Carbonate Dissolving Waters on Chemical and $^{13}C/^{12}C$ Compositions

P. Staniaszek, S. Halas

(Institute of Physics, Maria Curie-Sklodowska University [1])

The models for the chemical evolution of carbonate waters, developed by Deines et al. (1974), have been extended to include the effects of mixing on molalities of individual species. These effects, which mainly result from the redistribution of carbon-bearing species in the mixture, are strongly nonlinear and asymmetric with respect to the fraction of one solution in a binary mixture. The algorithm of the calculation and some selected examples of the mixing effects are discussed.

Die Modelle der chemischen Entwicklung von Karbonatwässern, wie sie von Deines et al. 1974 vorgeschlagen wurden, sind auf Mischungseffekte individueller Spezies erweitert worden. Diese Effekte, die hauptsächlich auf eine Neuverteilung von kohlenstoffhaltigen Spezies in der Mischung zurückgeführt werden, sind nichtlinear und asymmetrisch in bezug auf den Anteil einer binären Mischung. Der Berechnungsalgorithmus und einige ausgewählte Beispiele des Mischungseffektes werden diskutiert.

Keywords

calcite; carbon 13; carbonates; carbon dioxide; dolomite; electrolytes; ground water; isotope ratio

Introduction

An extensive study of carbonate dissolving waters in limestone terrains of the SW Lublin Upland, Poland, were carried out in 1977 [1, 2]. This project included observations of the activity of hydrogen ions ($aH^+ = 10^{-pH}$), molality of bicarbonate ions ($mHCO_3^-$) and $^{13}C/^{12}C$ isotope ratio of total inorganic carbon ($\delta^{13}CT$) in waters periodically sampled from 18 representative springs and 6 wells. The results obtained were compared with simple geochemical models developed by *Deines* et al. [3]. Since in some instances quite large departures from these models were noticed, we attempted to explain relatively rapid shifts in pH, $mHCO_3^-$ and $\delta^{13}CT$ by the mixing effects of two kinds of carbonate waters.

A rigorous calculation of the chemical composition of two solutions is a difficult task involving many experimental data on chemical equilibria and the appropriate *Debye-Hückel* expression. Therefore a search for a simplified approach is highly desirable. A common procedure in the calculation of water chemistry is to use the mass and the charge balance equations (cf. [4]). In case of carbonate groundwaters, of low total mineralization even at saturation with calcite and/or dolomite, further simplifications in the calculation procedure are justified. *Deines* et al. [3] in their calculation of carbonate water chemistry used an approach in which the initial pH value of a water (before carbonate dissolution) was assumed to vary independently of the initial partial pressure of CO_2 (PCO_2). In consequence the charge balance equations involving Ca^{2+}, Mg^{2+}, HCO_3^-, CO_3^{2-}, H^+ and OH^- may no longer be used since other ions are assumed to be present in unspecified amounts. Insofar as these ions contribute to ionic strength (I) and so influence activity coefficients (γ values), this assumption weakens the models. However, a comparison of the activity coefficients computed by this method, with those determined from total analyses of the investigated carbonate ground waters shows differences of only a few per cent or less between them. Thus the models, developed by *Deines* et al. [3], approach natural conditions quite colsely despite their simplicity. Of course, more sophisticated and complicated models for the chemistry of a water in a hydrologic system have been developed since 1974 [4–12]. However, the models of *Deines* et al. [3] still seem to be reliable for interpretation of real field data regarding the chemistry and $^{13}C/^{12}C$ ratios of the total inorganic carbon dissolved, if extention towards mixing of two carbonate waters will be made.

Mixing of carbonate waters have been considered by *Wigley* and *Plummer* [13] on the bases of particular form of the charge balance equation to avoid having to balance H^+ which is computationally inefficient. Consequently H^+ is calculated indirectly from total alkalinity and total carbon. The authors pointed out nonlinear behaviour of calcite saturation index (SI_c) and pH value of a binary mixture with respect to the fraction of the component being admixed.

It is the purpose of this paper to demonstrate the mixing effects on the concentration of the individual species $mHCO_3^-$, SI_c, pH versus fraction of one solution in the mixture. Finally, modifications in the evolution path developed by *Deines* et al. [3] regarding $mHCO_3^-$ and $\delta^{13}C$ due to the mixing effects will be considered. Since we will work within the frame-work outlined by these authors, the results obtained below are valid for typical ground waters occuring in carbonate aquifers being dissolved by soil-generated CO_2 transported in the infiltrating water.

Assumptions and definitions

The initial compositions, i.e. before mixing, are calculated for different stages of chemical evolution of both waters. For simplicity we consider the case in which the waters to be mixed evolved under closed system conditions (CSC).

[1]) Address: 20-031 Lublin, Poland

Tab. 1. Interpolated values of equilibrium constants at $T = 9{\cdot}4$ °C

Reaction	Symbol	Value	Ref.*)
$H_2O = H^+ + OH^-$	K_W	$0{\cdot}276 \cdot 10^{-14}$	[19]
$CO_2 + H_2O = H_2CO_3$	K_{CO_2}	0·05488	[20]**)
$H_2CO_3 = H^+ + HCO_3^-$	K_1	$3{\cdot}379 \cdot 10^{-7}$	[20]**)
$HCO_3^- = H^+ + CO_3^{2-}$	K_2	$3{\cdot}193 \cdot 10^{-11}$	[21]
$CaCO_3 = Ca^{2+} + CO_3^{2-}$	K_c	$4{\cdot}392 \cdot 10^{-9}$	[22]

*) New and improved data are recently available (Plummer and Busenber, [17]).

**) here $H_2CO_3 = CO_2aq + H_2CO_3$

The considerations included open system conditions (OSC) lead to similar results.

The term OSC refers to a permanent contact of the water with a large homogeneous reservoir of CO_2 under constant PCO_{2res}, whereas under CSC it is assumed that the water is isolated from the reservoir during carbonate dissolution [14].

It must be emphasized that the chemical composition of waters to be mixed (Tab. 2) is not assumed but calculated by using the models developed by *Deines* et al. [3]. As a consequence of using those models we get electrically unbalanced solutions though their major species are quite precisely determined. The obvious difficulties which arise with this "artificial" electric charge in the chemistry calculation of a mixture may be avoided as follows:

1) In order make these calculations the starting solutions must be perfectly balanced. To achieve this balance one must add or substract charge using a chemically inactive species (such as Cl^- or K^+). This is the procedure adopted by *Plummer* et al. [5] and *Plummer* [7]. It is almost necessary when using real field data.
2) The charge conservation equation may be applied to this apparent charge, the specific value of which (in coulombs/kg) is proportional to

$$R = 2\,mCa^{2+} + mH^+ - (2\,mCO_3^{2-} + mHCO_3^- + mOH^-) \tag{1}$$

The second way will be used in the calculation presented below. Formula (1) is written for calcite dissolution and it should be replaced by a similar formula in case of dolomite dissolution.

Calculation procedure

Let R_1 and R_2 denote the specific charge of the solutions being mixed. Thus R of the mixture may be expressed as linear combination of R-values of the end-members

$$R = xR_1 + (1 - x)\,R_2 \tag{2}$$

where x is the fraction of solution 1 in the mixture.

The conservation principle provides three further equations of similar character:

$$CT = xCT_1 + (1 - x)\,CT_2 \tag{3}$$

$$mCa^{2+} = xm_1Ca^{2+} + (1 - x)\,m_2Ca^{2+} \quad \text{and} \tag{4}$$

$$\delta^{13}CT = x\delta_1^{13}CT + (1 - x)\,\delta_2^{13}CT \tag{5}$$

where

$$CT = mH_2CO_3 + mHCO_3^- + mCO_3^{2-} \tag{6}$$

denotes carbon total in a solution and $\delta^{13}CT$ is considered a value to be proportional to the concentration of ^{13}C isotope in carbon total, which is the case for rather a narrow range of $\delta^{13}C$ variation in all relevant natural conditions.

The calculation of the chemical composition of a mixture starts from an evaluation of the activity of hydrogen ions aH^+. For this goal the expressions for CT and R values of the mixture are rewritten using respective equilibrium constants, given in Tab. 1, as follows

$$CT = mHCO_3^-\left(1 + \frac{mH_2CO_3}{mHCO_3^-} + \frac{mCO_3^{2-}}{mHCO_3^-}\right) = mHCO_3^-\left(1 + \frac{\gamma_1 aH^+}{K_1} + \frac{\gamma_1 K_2}{\gamma_2 aH^+}\right) \tag{7}$$

$$R = 2mCa^{2+} + aH^+/\gamma_1 - mHCO_3^- \times \left(1 + \frac{2\gamma_1 K_2}{\gamma_2 aH^+}\right) - \frac{K_W}{\gamma_1 aH^+} \tag{8}$$

where γ_1 stands for activity coefficients of all single charged ions, while γ_2 is the activity coefficient of CO_3^{2-}.

By eliminating $mHCO_3^-$ from the above equations, one obtains the following equation for $y = aH^+$:

$$y^4 + Ay^3 + By^2 + Cy + D = 0 \tag{9}$$

where

$$A = K_1/\gamma_1 + (2mCa^{2+} - R)\,\gamma_1$$

$$B = K_1(2mCa^{2+} - R + K_2/\gamma_2 - CT) - K_W$$

$$C = K_1\,[\gamma_1 K_2(2mCa^{2+} - R - 2CT)/\gamma_2 - K_W/\gamma_1]$$

$$D = K_1 K_2 K_W/\gamma_2$$

The appropriate real root of this polynomial can be computed by the Newton iterative method (e.g. [15]), using the weighted average of pH-values in the first step of the calculation. Actually further simplifications are justified when the calculation is limited to the pH ranging from 5 to 8. In this range of pH values, for a typical carbonate water ($2mCa^{2+} \approx mHCO_3^- \approx 0{\cdot}01$ mol/kg), term D in equation (9) and some contributions of A, B and C coefficients may be safely neglected.

Tab. 2. The chemical composition of CSC model solutions*) used in mixing calculations. The molalities of HCO_3^-, and Ca^{2+} are expressed in ppm while the ionic strength, $I = {}^1/_2(m\,HCO_3^- + mH^+ + mOH^-) + 2\,(mCO_3^{2-} + mCa^{2+})$, was calculated by substituting the molalities in mmoles per 1000 g of the solution. The partial pressure of CO_2 for waters which evolved under CSC was calculated: $PCO_2 = mH_2CO_3/K_{CO_2}$.

No.	$-\log \frac{PCO_{2\,res}}{atm}$	$-\log \frac{PCO_2}{atm}$	pH	HCO_3^- ppm	Ca^{2+} ppm	I	SI
1	·7	97	6·67	645	204	0·155	0
2	1·0	1·43	6·97	432	138	·0104	0
3	1·3	1·99	7·33	272	88	·0066	0
4	1·5	2·42	7·61	190	61	·0046	0
5	1·7	2·91	7·94	128	41	·0031	0
6	1·0	1·20	6·50	250	78	·0060	−1·02
7	1·3	1·57	6·70	162	51	·0039	−1·14
8	1·0	1·10	6·20	151	46	·0035	−1·71
9	1·5	1·63	6·20	47	14	·0011	−2·63

*) Computed according to the CSC model developed by *Deines* et al. [3] assuming the initial pH value before carbonate dissolution = 5·0 for each water

Thus a considerably simplified equation

$$y^3 + [(2m\mathrm{Ca}^{2+} - R)\,\gamma_1 + K_1/\gamma_1]\,y^2 + K_1(2m\mathrm{Ca}^{2+} - R - \mathrm{CT})\,y + K_1K_2\gamma_1(2m\mathrm{Ca}^{2+} - R - 2\mathrm{CT})/\gamma_2 = 0 \qquad (10)$$

is to be solved. The last term in equation (10) remains significant only if pH approaches 8.0.

Once $a\mathrm{H}^+$ is evaluated the $m\mathrm{HCO}_3^-$ of the mixture is calculated from equations (3) and (7). Finally,

$$m\mathrm{CO}_3^{2-} = \frac{\gamma_1 K_2}{\gamma_2 a\mathrm{H}^+}\, m\mathrm{HCO}_3^- \qquad (11)$$

and calcite saturation index

$$SI_c = \log \frac{\gamma\mathrm{Ca}^{2+}\, m\mathrm{Ca}^{2+}\, \gamma_2 m\mathrm{CO}_3^{2-}}{K_c} \qquad (12)$$

are calculated.

The molalities of the major species in the mixture are calculated iteratively by taking γ-values to be equal to 1 in the first step. Then with approximate molalities the ionic strength is estimated and new γ-values are evaluated from complete *Debye-Hückel* expression (cf. [16, 17]).

Results and discussion

To demonstrate the applicability of the algorithm derived we shall consider several mixing examples. All the calculations below are performed for 9·4 °C, the average temperature of underground waters studied in the Lublin Upland [1]. In $^{13}\mathrm{C}/^{12}\mathrm{C}$ calculation $\delta^{13}\mathrm{C}$ in the carbonates to be dissolved is assumed to be zero which is close to those of the Lublin Upland [18]. The calculations are performed on HP-85 computer.

The chemical characteristics of the model waters to be mixed are presented in Tab. 2. A wide differentiation in chemical composition of those waters allows a modelling of the changes in water chemistry during mixing.

In the present mixing calculation large departures were noticed of the resulting value of the mixture from the linear combination of the values of the end-member solutions. Such nonlinear effects are predominantly attributed to any or all of the following factors [13]:

1) redistribution of carbon-bearing species, the $\Delta P\mathrm{CO}_2$ effect,
2) purely algebraic effect, and
3) the nonlinear dependence of activity coefficients on ionic strength.

It should be emphasized that in many cases the curves presented below are highly asymmetric with respect to fraction of solution 1, denoted by x. Waters saturated under higher $P\mathrm{CO}_2$ are called solution 1.

Fig. 1 a-b shows the nonlinear effects of the closed system mixing on saturation index of the mixture for selected pairs of saturated waters ($SI = 0$). The dominant factor yielding undersaturation of the mixture is the difference in $P\mathrm{CO}_2$ between solutions prior to mixing. This factor significantly exceeds the algebraic effect, producing some increase in SI of the mixture which, in turn, exceeds the effect of nonlinear dependence of the activity coefficients on ionic strength. Consider, for example, pairs 4–1, 5–2, of the model waters all differing in $P\mathrm{CO}_2$ by the same factor ($10^{1.4}$). It is clear that the mixing effects on SI are almost identical, even though the various pairs differ significantly in ionic strength.

All the curves in Fig. 1 a-b have minima at a small fraction of the water of higher $P\mathrm{CO}_2$. For pair 5–1, for instance, the minimum value of SI is $-0{\cdot}56$ at $x = 0{\cdot}1$. The curves tend to symmetry for small differences in $P\mathrm{CO}_2$, but the minima always occur at $x < 0{\cdot}5$. Therefore, in natural conditions serious changes in water chemistry may occur due to admixture of a small fraction of water which was saturated at higher $P\mathrm{CO}_2$ than that of the original water.

In Fig. 1 c two examples of the mixing effects are shown in case when the second water is undersaturated ($SI < 0$). Again, the differences in $P\mathrm{CO}_2$ cause significant nonlinearities of a similar character as was the case of two saturated solutions. However, as pair 4–9 shows, the algebraic effect becomes significant when the difference in $P\mathrm{CO}_2$ is small, but there are large differences in the activities of Ca^{2+}.

The curves of pH as a function of x, plotted in Fig. 2 for different pairs of waters, show that at high difference in $P\mathrm{CO}_2$ the variations are highly nonlinear. A small fraction

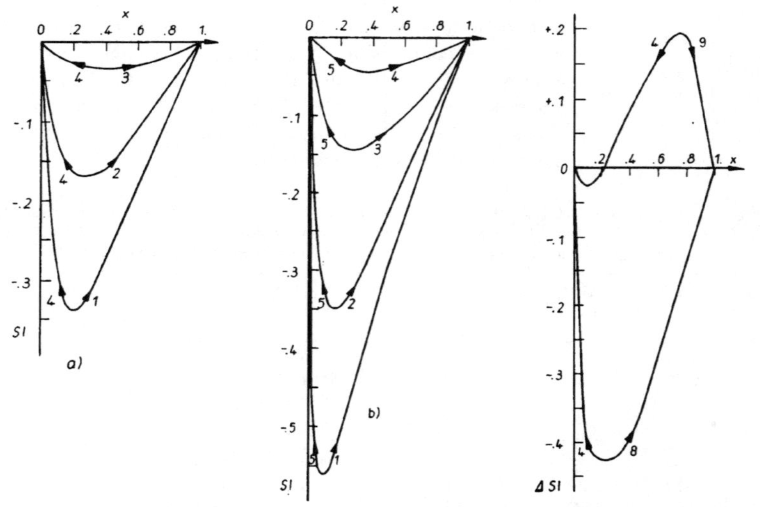

Fig. 1. a – b Calcite saturation index of the mixture of selected pairs of saturated ($SI = 0$) solutions from Tab. 2 as a function of solution 1 in the mixture

waters saturated unter higher $P\mathrm{CO}_2$ are called solution 1 c). Nonlinear effect on SI of the mixture in case when water 1 is undersaturated ($SI < 0$), $\Delta SI = SI_{\mathrm{mixture}} - xSI_{\mathrm{sol.\,1}} - (1 - x)\,SI_{\mathrm{sol.\,2}}$

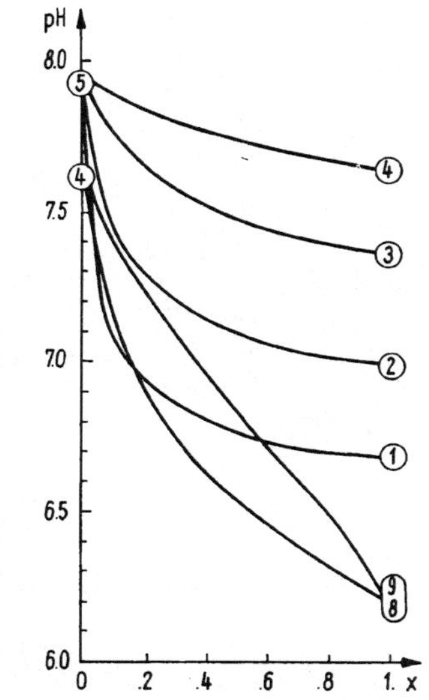

Fig. 2. Plots of pH value versus x for mixtures of two solutions selected from Tab. 2

of water evolved at a higher PCO_2 mixed with a lower PCO_2 water causes a strong reduction in pH of the mixture when compared with that of the original water. For an illustration of this effect, let us consider pair 5—1. Fig. 2 shows that the addition of fraction 0·05 of solution 1 to solution 5 changes the pH value from 7·94 to 7·30. If the fraction of solution 1 increases to 0·15 then the pH drops to 7·00, while at $x = 0{\cdot}25$ the pH attains 6·88 (compare this with the final value 6·67 at $x = 1$).

High sensitivity of pH to the addition of a small fraction fo CO_2-rich component is also reflected in the combined curves HCO_3^- vs pH and $\delta^{13}CT$ vs pH, plotted in Fig. 3 and 4. On the other hand, the calculation shows that changes in $mHCO_3^-$ of the mixture as a function of x are highly linear with x.

The combined mixing curves for $mHCO_3^-$ vs pH with x values marked along the curves are plotted in Fig. 3. These curves have been plotted on the background of selected carbonate-water evolution lines of *Deines* et al. [3]. Let us consider the following two extreme cases of closed-system mixing.

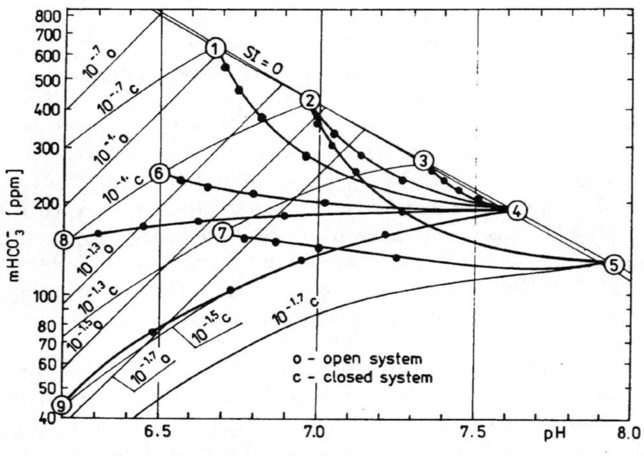

Fig. 3. Combined curves $mHCO_3^-$ versus pH for mixtures of two waters (from Tab. 2) plotted along the lines of calcium carbonate dissolution model of *Deines* et al. [3]

The mixing curves joining initial compositions of two waters are marked by dots for every. 2 change in x value

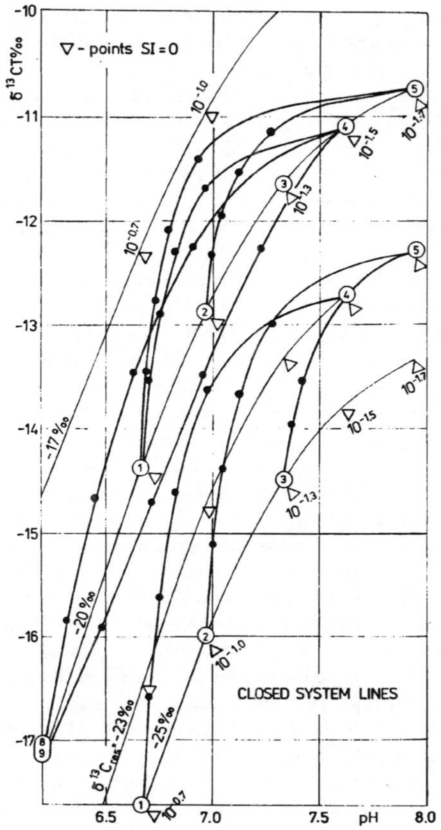

Fig. 4. Combined plots $\delta^{13}CT$ of the mixture versus pH on the background of the model evolution lines (*Deines* et al., [3])

The mixing curves joining initial compositions of two waters are marked by dots for every. 2 change in x value. The initial pH values are taken from Tab. 2, while $\delta^{13}CT$ were calculated assuming $\delta^{13}C_{res} = -20$, -23 and -25 ‰. Triangles show points of $SI = 0$ on the evolution lines for PCO_2 specified

1) A saturated water ($SI = 0$) is admixed with other saturated water which has a higher PCO_2 or undersaturated one evolving under a higher PCO_2.
2) A highly undersaturated water is admixed with a small fraction of a saturated water, which evolved under the same PCO_2.

In the first case, since pH is highly sensitive to the mixing process, an admixing can be easily recognized, while in the second case a point representing pH and $mHCO_3^-$ of the water to be examined will move along a closed system evolution line (e.g. water 9 admixed with water 4, Fig. 3). Thus mixing will be difficult to be recognized from pH and HCO_3^- data. The second case, however, is less likely in natural conditions. We shall see that the use of $\delta^{13}CT$ in such a case may allow to distinguish between carbonate dissolution and the mixing process.

Fig. 4 shows the combined mixing curves for $\delta^{13}CT$ vs pH along with some selected carbonate-water evolution lines calculated according to the simplified model of *Deines* et al. [3]. The parameter x varies along the mixing curves, and it is marked for every 0·2 change in the fractional composition of the mixture. When comparing the mixing curves with those describing the evolution of a single water under certain conditions (OSC, CSC, different $\delta^{13}C_{res}$) it is clear that the mixing effects can drastically change both pH and $\delta^{13}CT$ due to admixture of a small fraction of water different from that accumulated in the aquifer. Thus $\delta^{13}C_{res}$ derived under the assumption of no mixing by backtracking along a closed or open system path may be substantially in error.

If the aquifer water was saturated at low PCO_2 and subsequently admixed with a water saturated (or undersaturated) at a higher PCO_2, but at the same $\delta^{13}C_{res}$, then a major change in $\delta^{13}CT$ and pH will occur. Such a case is shown in Fig. 4, pairs 4-1, 4-8 and 4-9. More significant change will take place when the water to be admixed differs in $\delta^{13}C_{res}$ (consider mixing line 4-1 joining lines $\delta^{13}C_{res} = -23$ and -25 per mill, for example).

In a reverse situation, i.e. when a small fraction of low PCO_2 water is admixed with a CO_2-rich water, one can predominantly distinguish between carbonate dissolution and the mixing process. Let us consider, for example, the mixing lines 1 – 4 in Fig. 4. The line which starts and terminates on the evolution line for $\delta^{13}C_{res} = -20‰$ deviates from this line a similar way as that joining the evolution lines for $\delta^{13}C_{res} = -25$ and -23 per mill. In rather rare cases, however, the change in pH and $\delta^{13}CT$ can tend along the evolution line. Such a situation is anticipated, for example, in case of water, the chemistry of which is somewhere between that of waters 8 and 9. It should be noted that the mixing curve lies below the evolution line when the algebraic effect exceedes the ΔPCO_2 effect [13]. Therefore, if these effects compensate each other, then the changes in pH and $\delta^{13}CT$ will move along the evolution line and, thus, the mixing process cannot be distinguished from the carbonate dissolution.

The situation when both processes may be indistinguishable may occur rather seldom because natural variation of $\delta^{13}C_{res}$ diminishes the probability that the mixed waters evolved under identical $\delta^{13}C_{res}$. Thus the mixing process may be frequently recognized from $\delta^{13}CT$ and pH though $mHCO_3^-$ and pH data are insufficient for this purpose.

Although the mixing of carbonate waters in natural conditions is frequently a continuous process, which further complicates mathematical description, the simplified model presented here may be particularly useful in the studies of infiltration time. This inference is drawn from the fact that, due to the nonlinear effects, even a small discrete inflow into an aquifer can significantly change pH and $mHCO_3^-$ and Ca^{2+} without evident changes in cation chemistry The most remarcable changes in pH and $mHCO_3^-$ are expected when the inflowing water has a partial pressure of CO_2 higher than the water accumulated in the aquifer. Such a situation frequently occurs during spring when melting and/or rainfall waters start to feed the underground reservoirs.

The use of the $\delta^{13}CT$ criterion for studies of infiltration time is particularly useful when the admixed water has almost identical evolution path like that accumulated in the aquifer. In such case, if both waters are already saturated ($SI = 0$), the chemistry of the mixture will not change at all. The $\delta^{13}CT$ value, however, may be affected due to the variation in $^{13}C/^{12}C$ of carbon dioxide generated in the soil.

Mixing evidences on real field data

To illustrate the usefulness of the mixing criteria, we shall consider two examples taken from the work by *Halas* et al. [1, 2]. Fig. 5 shows changes in $mHCO_3^-$ and $\delta^{13}CT$ vs pH observed during a period from January 26 to May 5, 1977. The samples of water were taken in the area of the SW Lublin Upland in which Crateceous (spring no. 1) and Tertiary (spring no. 13) limestones are covered by Plei-

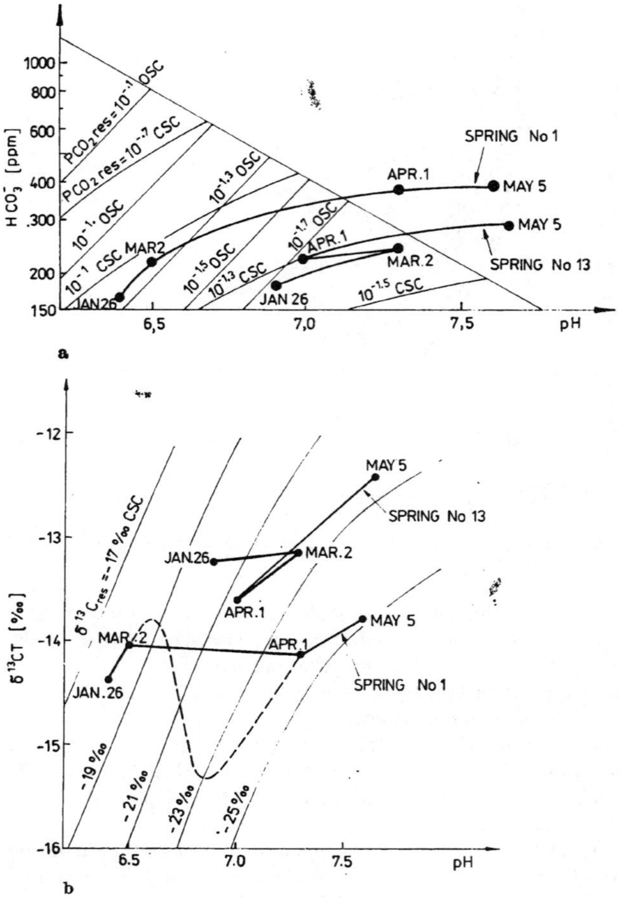

Fig. 5. a – Variations in $mHCO_3^-$ versus pH and b – in $\delta^{13}CT$ versus pH along the evolution lines calculated according to *Deines* et al. [3]

Both examples represent natural variations which were recorded by *Halas* et al. [1] in two springs located in SW Lublin Upland

stocene eolic and soilfluctional rocks of variable thickness. The period of observations was very advantageous for studies of infiltration rate since numerous inflow events were feeding the limestone aquifer in the late autumn, winter and early spring, whereas the feeding was stopped in April by increased evaporation of rainfall water and by an increase of biological activity in the soil.

The waters sampled in January 26 were highly undersaturated as indicated by the initial points in Fig. 5. The main feature of the curves presented in Fig. 5 (as well as of those not presented in this figure) is a strong variation in chemistry and $\delta^{13}CT$ during a period from January 26 to April 1, and a common trend of those values from April 1 to May 5. This trend towards saturation nearly approximates the evolution lines calculated according to *Deines* et al. [3]. The variations observed reflect the mixing processes which in natural conditions are accompanied by carbonate dissolution.

It should be emphasized that the mixing events may be unrecognized from $mHCO_3^-$ and pH data. In case of spring no. 1 the points in Fig. 5a lie within experimental errors on the evolution path. The mixing process, however, is clearly visible from $\delta^{13}CT$ and pH data in Fig. 5b. A small change in $mHCO_3^-$, $\delta^{13}CT$ and pH, which occured between January and March indicates inflows of water having similar PCO_2 and $\delta^{13}CT$. These inflows apparently moderated the carbonate dissolution. However, in March, the shift in Fig. 5a indicates the carbonate dissolution, whereas that in Fig. 5b shows the admixture of water carrying inorganic carbon from CO_2-reservoir being significantly lowered in $\delta^{13}C$ (see Fig. 4). A presumable path of $\delta^{13}CT$ vs pH during this period is shown by the dashed line. In case of spring no. 13 the changes recorded in February were similar to those observed for spring no. 1 in March. During March a significant inflow of CO_2-rich water was recorded in Figs. 5a and 5b. During April, further changes indicate only the carbonate dissolution process.

Acknowledgements – This work was supported by MR. I. 5 Program, University of Warsaw. We are grateful to Dr. *T. M. L. Wigley* for comments on the manuscript.

Received January 7, 1985
Accepted in revised form April 15, 1985

References

[1] *S. Halas, B. Janiec, P. Staniaszek,* The investigation of variation reasons of $^{13}C/^{12}C$, pH and concentration of inorganic carbon in underground waters (*in Russian*). IX International Symposium on Stable Isotopes in Geochemistry. Moscow, 16–19 November 1982, pp. 447–449

[2] *S. Halas, B. Janiec, P. Staniaszek,* Seasonal variations in chemistry and $^{13}C/^{12}C$ ratios of carbonate dissolving water. International Symposium on Isotope Hydrology in Water Resources Development, Vienna, 12–16 September 1983. The book of extended synopses, p. 174. The extended version of this paper has been submitted to Nordic Hydrology.

[3] *P. Deines, D. Langmuir, R. S. Harmon,* Geochim. Cosmochim. Acta **38** (1974) 1147

[4] *L. N. Plummer, D. L. Parkhurst, D. C. Thorstenson,* Geochim. Cosmochim. Acta **47** (1983) 665

[5] *L. N. Plummer, D. L. Parkhurst, D. R. Kosiur,* MIX 2: A computer program for modelling chemical reactions in natural waters U. S. Geol. Survey, Water Resources Inv., 75-61 (1975), 68 p. (U. S. Dept. Commerce, NTIS, Springfield, VA 22161, Accession No. PB251668)

[6] *L. N. Plummer, B. F. Jones, A. H. Truesdell,* WATEQF – A FORTRAN IV version of WATEQ, a computer program for calculating chemical equilibrium of natural waters. U. S. Geol. Survey, Water Resources Inv. 76-13, (1976) 61 p. (U. S. Dept. Commerce, NTIS, Springfield, VA 22161, Accession No. PB261027)

[7] *L. N. Plummer,* Water Resources Res. **13** (1977) 801

[8] *T. M. L. Wigley, L. N. Plummer, F. J. Pearson* Jr., Geochim. Cosmochim. Acta **42** (1978) 1117

[9] *T. M. L. Wigley, L. N. Plummer, F. J. Pearson* Jr., Errata, Geochim. Cosmochim. Acta **43** (1979) 1395

[10] *L. N. Plummer, W. Back,* Amer. J. Sci. **280** (1980) 130

[11] *D. L. Parkhurst, D. C. Thorstenson, L. N. Plummer,* PHREEQE – A computer program for geochemical calculations. U. S. Geol. Survey, Water Resources Inv. 80–96 (1980) 210 p., (U. S. Dept. Commerce, NTIS, Springfield, VA 22161, Accession No PB81-167801)

[12] *D. L. Parkhurst, L. N. Plummer, D. C. Thorstenson,* BALANCE – A computer program for calculation of chemical mass balance. U. S. Geol. Survey, Water Resources Inv. 82–14 (1982) 33 p. (U. S. Dept. Commerce, NTIS, Springfield, VA 22161, Accession No PB82-255902)

[13] *T. M. L. Wigley, L. N. Plummer,* Geochim. Cosmochim. Acta **40** (1976) 989

[14] *D. Langmuir,* Geochim. Cosmochim. Acta **35** (1971) 1023

[15] *P. A. Stark,* Introduction to numerical methods. London: MacMillan Co. 1970

[16] *A.H. Truesdell, B. F. Jones,* U. S. Geol. Survey Research **2** (1974) 233

[17] *L. N. Plummer, E. Busenberg,* Geochim. Cosmochim. Acta **46** (1982) 1011

[18] *S. Halas, J. Lis, J. Szaran, A. Trembaczowski, W. Wolacewicz, W. Zuk,* Przeglad Geologiczny **27** (1979) 162 (*in Polish*)

[19] *T. H. Ackermann,* Z. Elektrochem. **62** (1958) 411

[20] *H. S. Harned, R. Davis,* Jr., J. Amer. Chem. Soc. **65** (1943) 2030

[21] *H. S. Harned, S. R. Scholes* Jr., J. Amer. Chem. Soc. **63** (1941) 1706

[22] *R. L. Jacobson, D. Langmuir,* Geochim. Cosmochim. Acta **38** (1974) 301

Isotopenpraxis **21** (1985) 12, S. 429 bis 432

Определение коэффициентов перехода в системе почва-растение в окрестностях АЭС

Т. Сабова, Ш. Барта, В. Стано, И. Шмидт, М. Калинчак

(Институт радиоэкологии и использования ядерной техники, Кошице)[1]

В 1981—82 гг отбирались образцы почв и растений в месте строительства АЭС, которые использовалис для определения коэффициентов перехода в системе почва-растение. Полученные значения концентрацион ных факторов колеблются в зависимости от типа растения и от типа почвы. Растения, у которых для анализов отбиралась надземная часть (особенно кормовые) характеризуются более высокими значениями концентрационных факторов по сравнению с растениями, у которых анализировалась непосредственно потребляемая часть растения. Значение концентрационного фактора для стронция определяется типом почвы, реакцией почвы, обменной емкостью и содержанием обменного кальция.

In the years 1981 and 1982 samples of soil and plants from the localities of future nuclear power plant buildings were investigated on transfer-coefficients in the soil-plant system. The concentration factors vary depending on plant species and type of soil. The concentration factors are higher in pasture plants, where the overground part of the was analysed, than in plants, where the analysed part of the plant is important for human consumption. The concentration factor for strontium depends on soil type, exchange capacity and the content of exchangable calcium in the soil.

Keywords

fallout deposits; food; ion exchange; plants; radionuclide migration; soils; terrestrial ecosystems

Оценку радиационной экспозиции населения окрестностей атомноэнергетических оборудований, расчеты концентраций радионуклидов в звеньях окружающей среды, основанные на обьеме выбросов, должны проводиться как во время строительства, так и во время эксплуатации оборудования. Для этих

[1]) Адрес: ЧССР, 040 61 Кошице, Коменского 9

расчетов важно значение коэффициентов перехода для звеньев пищевой цепочки.

Поступление радионуклида из почвы в растения характеризуется фактором перехода $\mathcal{ФП}_{ПР}$, последний определяется отношением концентраций нуклида в растении и в почве [1—4].

Значения факторов перехода, исппользуемые в расчетных моде лях, переняты из зарубежной литературы, в связи с этим необходимо проверить их в экологических условиях ЧССР. К факторам среды, влияющим на факторы перехода, относятся характеристики почв, тип растения, время года, часть растения, отбираемая для образцов [3, 5].

Целью настоящей работы является определение коэффициентов перехода в системе почва-растение для окрестностей строительства АЭС, в зависимости от типа почвы и типа растений.

Материал и методика

В 1981—82 гг отбирались образцы почвирастений в окрестностях строительства АЭС. В отобранных образцах почв были определены агрохимические характеристики.

В образцах почв и растений было определено содержание экстрагируемых кальция, стронция, хрома, меди, железа, марганца, натрия, никеля и цинка.

Почвы высушивались при 105 °С до достижения постоянного веса. 10 г сухой почвы экстрагировалось в 50 мл 1 М HCl в течение 1 часа, при постоянном перемешивании. Экстрат был профильтрирован и использован для определения отдельных элементов.

Растения высушивались при 105 °С до достижения постоянного веса. Для дальнейшего исследования отбиралось 50 г сухого образца. Образец растений сжигался в муфельной печи при температуре 550 °С, зола растворялась в смеси концентрированной HCl и HNO_3 в тепле. Во время растворения к образцу был 2 раза добавлен 1 мл 30% H_2O_2. Образец отпаривался до обьема 1—2 мл, после остывания переливался в колбу и водой доводился до 25 мл.

В приготовленных таким способом образцах определялись отдельные элементы следующими методами:

Cr — эмиссионная спектрометрия пламени при 422,7 нм

Sr — эмиссионная спектрометрия пламени при 460,8 нм

Na — эмиссионная спектрометрия пламени при 589,2 нм

Cr — ААС при 357,9 нм в воздушно-ацетиленовом пламени

Табл. 1. Агрохимические характеристики изученных почв окрестностей АЭС Моховце

Почва	pH/KCl	Гумус %	*V* %	T	S	H	Ca^{2+}	Содержание фракций, %			
				мвал/100 г				0,01 мм	0,001 мм	0,001 – 0,01 мм	0,01 – 0,05 мм
Бурая псеудоподзолистая	4,71	1,89	74,8	13,03	9,75	3,28	5,33	34,82	15,3	19,4	51,7
Бурая псеудоподзолистая	5,28	1,66	86,46	18,84	16,29	2,55	10,74	42,5	23,8	18,7	44,6
Бурая псеудоподзолистая	4,90	1,53	96,14	22,01	18,96	3,05	9,06	39,5	20,4	19,1	37,5
Бурая псеудоподзолистая	6,50	1,83	94,87	21,45	20,35	1,10	12,19	41,8	22,5	19,3	42,6
Бурая псеудоподзолистая	4,95	1,86	82,36	19,05	15,69	3,36	9,87	41,1	24,1	17,0	45,7
Бурая псеудоподзолистая	5,34	1,91	82,76	14,79	12,24	2,55	7,43	39,8	22,4	17,4	41,2
Бурая псеудоподзолистая	5,15	1,64	84,12	17,32	14,57	2,75	9,55	39,7	21,8	17,9	37,8
Бурая псеудоподзолистая	7,26	2,08	98,92	30,10	29,80	0,65	15,15	42,5	20,7	21,8	33,3
Бурая псеудоподзолистая	6,40	1,35	93,88	22,88	21,48	1,40	11,48	41,2	23,2	18,0	43,5
Бурая псеудоподзол. иллимеризованная	5,87	2,48	92,30	27,24	25,14	2,10	11,81	41,1	22,0	19,1	36,0
Бурая псеудоподзол. иллимеризованная	6,30	2,13	95,95	28,41	27,26	1,15	12,15	42,7	23,1	19,6	42,2
Бурая псеудоподзол. иллимеризованная	5,35	2,31	88,32	22,26	19,66	2,60	8,81	37,8	18,3	19,5	46,2
Чернозем оподзоленный	5,20	1,61	86,97	19,19	16,69	2,50	11,13	41,2	23,5	17,7	42,2
Чернозем оподзоленный	6,54	2,41	94,23	25,98	24,48	1,50	13,39	43,3	24,6	18,7	41,4
Чернозем оподзоленный	5,97	2,68	92,45	28,83	26,68	2,15	13,48	46,6	27,9	18,7	32,9
Аллювиальная	5,55	2,47	90,77	27,08	24,58	2,50	14,06	46,5	26,3	20,2	40,5
Аллювиальная	4,93	2,44	81,63	24,08	20,38	3,70	12,16	45,1	26,4	18,7	43,4

Табл. 2. Агрохимические характеристики изученных почв из окрестностей АЭС Дукованы

Почва	pH/KCl	Гумус, %	*V* %	T	S	H	Ca^{2+}	Содержание фракций, %			
				мвал/100 г				0,01 мм	0,001 мм	0,001 – 0,01 мм	0,01 – 0,05 мм
Бурая лесная	4,22	1,61	81,98	15,82	12,97	2,85	2,69	11,7	5,3	6,4	11,8
Бурая лесная	5,17	1,27	83,94	15,76	13,23	2,53	8,67	35,4	18,9	16,5	27,1
Бурая лесная	4,37	1,66	74,02	14,09	10,43	3,66	7,01	25,7	15,9	9,8	23,9
Чернозем	5,52	1,60	87,37	16,07	14,04	2,03	8,30	30,5	15,9	14,6	25,6
Бурая псеудоподзолистая	7,25	2,06	85,46	18,03	15,41	2,62	18,00	43,3	25,6	17,7	34,4
Бурая псеудоподзолистая на лессе	4,76	1,97	74,21	16,21	12,03	4,18	6,59	28,0	13,8	14,2	20,3
Бурая лесная	4,80	1,92	80,29	15,73	12,63	3,10	8,36	36,0	23,0	13,0	35,0
Бурая лесная	4,89	2,37	73,88	24,43	18,05	3,38	9,14	35,9	20,7	25,2	27,7
Бурая лесная	5,13	2,08	84,48	20,17	17,04	3,13	10,94	40,9	22,3	18,6	41,0
Чернозем	5,97	2,68	93,32	29,95	27,95	2,00	13,14	43,8	25,2	18,6	36,4
Бурая псеудоподзолистая	6,15	4,14	94,88	39,09	37,09	2,00	7,02	40,7	21,0	19,7	24,0
Бурая псеудоподзолистая	4,70	1,76	82,21	19,94	15,94	3,45	5,68	41,3	25,1	16,2	37,6
Луговая	5,70	8,39	90,9	34,81	31,36	3,45	4,90	27,7	13,5	14,2	16,2

Табл. 3. Определение коэффициентов перехода в системе почва-растение в окрестностях строительства АЭС

вид растения	МОХОВЦЕ										ДУКОВАНЫ									
	Cr	Cu	Fe	Mn	Na	Ni	Sr	Zn	Ca	D	Cr	Cu	Fe	Mn	Na	Ni	Sr	Zn	Ca	D
Ячмень-бурая лесная почва												1,03	0,04	0,07	1,8	0,07	0,11	8,81	0,22	0,590
Ячмень-чернозем												0,57	0,102	0,09	3,92	0,043	0,04	6,29	0,11	0,401
												1,01	0,19	0,07	3,69	0,13	0,11	3,59	0,26	0,505
Ячмень-чернозем оподзоленный		0,76	0,08	0,04	6,87		0,039	27,92	0,02	1,703										
Ячмень озимый-бурая лесная почва		0,84	0,03	0,05	12,5		0,11	13,86	0,29	0,41										
Пшеница-бурая лесная почва	0,77	0,92	0,3	0,19	5,88	0,39	0,29		1,59	0,184		1,32	0,37	0,13	1.52	0,085	0,13	9,87	0,57	0,236
												0,81	0,02	0,11	0,74	0,22	0,11	7,83	0,26	0,416
Пшеница-бурая лесная почва												1,2	0,32	0,28	1,55	0,64	0,365	17,56	0,58	0,733
Пшеница-бурая лесная почва		0,39		0,07	5,75		0,05	25,87	0,06	0,888										
Пшеница-бурая лесная почва на лессе				0,042	8,11		0,03	11,0	0,17	0,185										
Пшеница-чернозем												0,37	0,23	0,07	1,16	0,07	0,04	3,4	0,11	0,469
Пшеница-чернозем оподзоленный		0,43		0,05	1,5		0,059	24,05	0,11	0,555										
Пшеница-аллювиальная почва		0,09	0,06	0,08	5,63	0,24	0,07	23,75	0,05	0,661										
Люцерна-чернозем	0,43	1,54	0,35	0,018	54,29	0,9	4,4	31,46	6,55	0,670	1,25	3,36	0,53	0,36	23,14	1,25	7,99	47,06	10,81	0,869
Люцерна-бурая лесная почва	0,36	0,83	0,71	0,12	6,87	0,56	3,47	29,81	5,84	0,590	0,43	3,34	0,2	0,18	30,08	0,47	2,65	12,32	1,19	1,511
		1,3		0,09	63,25	0,44	3,23	24,72	5,68	0,574	0,34	1,9	0,28	0,24	21,19	1,16	4,52	9,3	2,15	2,10
1-ый покос	0,4	1,3	0,12	0,11	32,62	1,07	2,39	28,7	4,18	0,573										
2-ой покос		0,78	0,11	0,17	48,24	0,64	2,78	14,78	4,85	0,575	0,74	3,23	0,36	0,16	20,62	1,23	7,76	10,98	6,63	1,407
Люцерна-бурая лесная почва											0,31	1,24	0,09	0,15	13,19	0,74	6,53	6,78	9,20	0,714
											2,03	0,16	0,5	0,26	29,8	0,89	5,5	60,98	5,77	0,868
Люцерна-аллювиальная почва	0,26	0,79	0,15	0,19	16,8	0,53	2,32	18,34	4,16	0,565										
		0,48	0,11	0,007	20,6	0,44	2,81	19,36	4,36	0,653										
Люцерна-чернозем оподзоленный		0,86	0,02	0,17	69,35	0,63	3,96	29,81	5,39	0,743										
Клевер-бурая лесная почва		1,79	0,07	0,18	70,72	1,17	6,01	13,18	9,66	0,615										
		1,92	0,09	0,21	40,7	0,71	1,8	18,34	5,19	0,346										
Боб-бурая лесная иллимеризованная	0,86	1,75	0,45	0,19	224,0	1,15	2,74	26,06	3,41	0,807										
Луговая трава-луговая почва	0,66	0,05	0,17	0,09	25,12	0,3	0,77	10,58	0,76	1,052										
Перец-чернозем опод золенный		1,3	1,33	0,05	28,83	0,2	4,83	12,74	0,59	8,130										
Томат-чернозем оподзоленный	0,29	1,51	0,11	0,06	59,29	0,43	0,06	40,29	0,48	0,139										

Cu — ААС при 324,3 нм в воздушно-ацетиленовом пламени
Fe — ААС при 324,3 нм в воздушно-ацетиленовом пламени
Mn — ААС при 279,5 нм в воздушно-ацетиленовом пламени
Ni — ААС при 232,0 нм в воздушно-ацетиленовом пламени
Zn — ААС при 213,9 нм в воздушно-ацетиленовом пламени

Результаты и дискуссия

Агрохимические характеристики почв приведены в табл. 1 и 2.

Земля обеих местностей плодородна, с благоприятными для достижения высоких урожаев физико-химическими свойствами.

Коэффициенты перехода определялись для стабильных элементов тех радиоизотопов, которые представлены в выбросах, или относятся к коррозийным элементам. Все элементы приведены в табл. 3.

Коэффициенты перехода определялись по формуле:

$$ФП = \frac{\text{ppm элемента в свежей вегетации}}{\text{ppm элемента в почве}}$$

Более точным будет использование термина ,,концентрационный фактор,, (*КФ*), поскольку измерялась концентрация элемента в сухой вегетации и в сухой почве.

При сравнении обеих местностей были определены высокие концентрационные факторы для всех изучаемых элементов в случае люцерны, клевера, бобовых и сена из лугов.

Для зерна злаковых культур в случае обеих местностей отмечены наиболее низкие концентрационные факторы.

Изучая концентрационные факторы для стронция, одного из важных загрязнителей пищевой цепочки, приходим к подобной зависимости: наиболее высокие концентрации стронция получены для сена люцерны и клевера, зерна бобовых, а также для лугового сена, что согласуется с данными *Гоффмана* и *Келлера* [6], *Нг* и др. [1, 2] и *Рето Зах* [7]. Авторы отмечают повышенные концентрационные факторы для вегетативных органов по сравнению с генеративными органами. Наиболее низкие значения концентраций стронция определены для зерна злаковых культур, не была отмечена разница между озимовой и яровой формами культур.

Концентрационные факторы для стронция выше для образцов из окрестностей Дукованы, по сравнению с Моховцами, что может быть обьяснено различными свойствами почв, в первую очередь pH/KCl, обменной емкостью и содержанием обменного кальция [6].

Почвы из окрестностей Дукованы характеризуются более низкими значениями обменной емкости и более низким содержанием обменного кальция, что влечет за собой повышенное поступление стронция в многолетние кормовые, а также в зерно злаковых культур. Эти повышенные переходы указывают, что поступление стронция из почвы будет таким же в случае применения этих данных для долгоживущих нуклидов.

При поступлении радиостронция в растения осуществляется дискриминация его по отношению к кальцию. В большинстве случаев соотношение:

$$Д = \frac{^{90}\text{Sr/Ca в растении}}{^{90}\text{Sr/Ca в почве}}$$

меньше единицы [8, 9].

В соответствии с литературными данными можно констатировать, что фактор дискриминации достигает найвысших значений для виковых растений и лугового сена, а самых низких значений — для зерна злаковых культур.

Сравнение полученных нами данных с Regulatory Guide 1·109 обнаруживает, что определенные нами концентрационные факторы выше. Полученные результаты нужно дополнить данными о некоторых типах растений и учесть влияние экологических и почвенных факторов на определение концентрационных факторов для радиогигиенически значимых элементов.

Выводы

Исходя из полученных данных можно констатировать, что концентрационные факторы сильно изменяются в зависимости от типа растения и типа почвы. Растения, у которых анализу подвергались надземные части растений (особенно фуражные культуры), характеризуются более высокими значениями концентрационных факторов по сравнению с растениями, у которых анализировалась лишь непосредственно потребляемая часть растения. На значение концентрационных факторов влияют тип почвы, реакция почвы, обменная емкость и содержание обменного кальция.

Поступила в редакцию 6/II, 1985г

Литература

[1] *Y. C. Ng, C. S. Odsher, S. E. Thompson*, Soil-to-Plant Concentration Factors for Radiological Assessments, NUCID-19463, November 1982

[2] *Y. C. Ng, C. S. Colsher, S. E. Thompson*, Transfer coefficients for terrestrial Foodchain — their Derivation and Limitation, UCRL-81640, 1979

[3] *K. Heine, A. Wiechen*, Studies of the Transfer Factors of ^{90}Sr and ^{137}Cs in the Foodchain Soil-Plant-Milk, 5th International Congres of IRPA, Jerusalem, 9—14. March 1980, abstract No. 1305

[4] *Y. C. Ng*, Nuclear Safety **23** (1982) 57

[5] *J. D. Sartor, W. B. Allen, J. J. Allen*, Uptake of Rydionuclides by Plants, MW-5893, 1966

[6] *L. G. Hoffman, J. H. Keller*, Characterization of Soil to Plant Transfer Coefficients for Stable Cesium and Strontium, NUREG-CR-2495, September 1981

[7] *Reto/Zach*, Transfer Coefficients to terrestrial Food Products in equilibrum assessment Models for Nuclear Installations, AECL-6449, September 1980

[8] *S. A. Szabó*, Fizikai Folyoirat **6** (1977) 445

[9] *S. A. Szabó*, Radiochemical Radioanalytical Letters **4** (1979) 283

Kurznachrichten / Newlets

Kernenergie:
ERFAHRUNGEN UND TENDENZEN BEI DER ENTWICKLUNG VON SCHNELLEN REAKTOREN

Vom 22.-26.7.1985 fand in Lyon ein internationales Symposium ueber Erfahrungen und Tendenzen auf dem Gebiet der Entwicklung von Schnellreaktoren statt. Das Symposium wurde von der IAEA in Zusammenarbeit mit der franzoesischen Organisation Electricité de France und dem Commissariat à l' Energie Atomique veranstaltet. Ende des Jahres 1984 waren 345 Leistungsreaktoren in 26 Laendern in Betrieb, von denen ungefaehr 13% des in der ganzen Welt erzeugten Stromes geliefert wurde. In sechs Laendern ist der Anteil des durch Kernenergie erzeugten Stromes hoeher als 35%, und in Frankreich ist er auf 50% gestiegen. Die gegenwaertige Entwicklung dieser Energiequelle beruht jedoch auf thermischen Reaktoren, bei denen zur Stromerzeugung nur ein kleiner Teil ihres natuerlich vorkommenden Uranbrennstoffs gebraucht werden kann. Die Bodenschaetze an Uran, die auf der Welt zu wirtschaftlichen Preisen zur Verfuegung stehen, sind begrenzt. Obwohl ihr Energiegegenwert mit den Oel- und Erdgasquellen in der Welt unter Anwendung der gegenwaertigen Technologie (thermische Reaktoren, einmalige Verwendbarkeit) verglichen werden kann, kann Kernenergie nur als eine verhaeltnismaessig kurzfristige Energiequelle betrachtet werden, da die Uranvorkommen innerhalb weniger Jahrzehnte erschoepft sein werden. Schnelle Brutreaktoren koennen das U-238 (das 99,3% des natuerlich vorkommenden Urans ausmacht) wirksam in spaltbaren Brennstoff umwandeln und mehr spaltbares Material erzeugen als im Reaktor verbraucht wird. Dies ermoeglicht den Gebrauch nicht nur von spaltbarem U-235, sondern von fast jedem spaltbarem Material. Im Vergleich zu thermischen Reaktoren koennte bei schnellen Brutreaktoren die Energiemenge, die aus Uran gewonnen werden kann, um einen Faktor von ueber 50 erhoeht werden. Zusammen mit nuklearem "Brueten" bietet die Kernenergiespaltung deshalb eine fast unerschoepfliche Energieversorgungsquelle fuer die unmittelbare Zukunft. Bedeutende Erfahrungen wurden auf dem Gebiet der Brutreaktorenentwicklung gewonnen. Fuenfzehn schnelle Brutreaktoren befinden sich derzeit in Betrieb oder im Bau. Einer von ihnen ist der Super-Phenix, Industriegroesse, (Frankreich), mit einer Elektrizitaetserzeugung von 1200 MWe, der in der Naehe von Lyon in Betrieb steht. Sechs weitere befinden sich im Planungsstadium. Anhand von Arbeiten, die bei dem Symposium in Lyon diskutiert wurden, sollte der Fortschritt, der bisher auf dem Gebiet der Entwicklung von schnellen Brutreaktoren gemacht wurde, und die Tendenzen bezueglich Entwuerfe fuer zukuenftige schnelle Reaktoren, ueberprueft werden. (IAEA-Pressekommunique (1985)10)

Radioecology:
ENVIRONMENTAL RADIOACTIVITY STUDIED IN DUMFRIES AND GALLOWAY

The UKAEA's Harwell Laboratory is to carry out a study of environmental radioactivity in the Scottish coastal region of Dumfries and Galloway. The study, initially for one year, has been commissioned by the Department of the Environment (DOE) on behalf of the Scottish Development Department (SDD) to provide information on the levels of artificial radionuclides in the environment, and on the transfer of radioactivity from sea to land. The investigation is in addition to the routine radioactivity monitoring programmes carried out by the SDD, and it will provide further data for radiological assessments in the region. The study will involve sampling in coastal waters, intertidal areas and beaches, the collection samples of sea spray, and the continuous collection of airborne particulate and deposited material at coastal and inland monitoring stations. The results will be compared with the models for predicting the transfer of radionuclides which have been developed through studies elsewhere. The findings on radionuclide levels will be compared with the internationally accepted limits for radionuclides in the environment and with internationally recommended standards of radiological safety. (Atom 345 (1985) 6)

Nuclear Energy:
NUCLEAR GENERATION IN OECD

Nuclear electricity production in the OECD countries grew at a rate of 18.7 per cent in 1984, the largest rate of increase since 1977, according to an annual survey just completed by the Nuclear Energy Agency (NEA). Compiled from official submissions by OECD countries (which account for 80 per cent of the world's nuclear generating capacity) the survey constitutes a comprehensive record of developments and projections in nuclear electricity generation, installed capacities and associated fuel cycle services. All 13 OECD countries with nuclear programmes, with the exception of the Netherlands, increased their nuclear electricity generation in 1984. Four countries-France, the FRG, Spain and Sweden - reported a more than 30 per cent increase. Electricity generation during the 12 month period rose by 4.7 per cent, from 5100 TWh to 5340 TWh. The share of total OECD electricity generation met by nuclear power increased from 16 per cent to 18 per cent and now exceeds the share met by oil. Total installed nuclear capacity in the OECD area reached 182 GWe at the end of 1984, an increase of 15 per cent. The number of operable nuclear power plants rose to 264, with 21 new reactors becoming operable in 1984. An additional 112 new reactors with a capacity of 116 GWe were under construction during this period and 27 reactors with 28 GWe were in the planing stage. ("Nuclear Power and Fuel Cycle Data in OECD Member Countries 1985", NEA Public Information Office, 38 Boulevard Suchet, 75016 Paris, France). (Atom 345(1985)7)

Homogenkatalyse:
KATALYTISCHE SYNTHESE VON ORGANO-Li-UND -Mg-VERBINDUNGEN

Als eine der neuen Entwicklungen auf dem Gebiet der Homogenkatalyse gelten Katalysatoren, die die Umsetzung von 1-Alkenen zu Alkenyl-Li-Verbindungen und LiH sowie die Hydrierung von Li und Mg unter milden Bedingungen ermoeglichen. Katalytisch erzeugtes MgH_2 ist hochreaktiv; es laesst sich katalytisch an 1-Alkene zu Diorgano-Mg-Verbindungen addieren und kann u.a. zur Synthese von SiH_4 und "aktivem Mg" angewendet werden. Die H_2-Speicherung in Form von Metallhydriden ist ebenfalls moeglich. Die Hydrier-/Dehydrierversuche ergaben, dass sich katalytisch hergestelltes MgH_2 (mit und ohne Dotierung mit Fremdmetallen) als Hochtemperatur-H_2-Speicher eignet. In einer Uebersichtsarbeit von B. Bogdanovic (Katalytische Synthese von Organo-Li- und Mg-Verbindungen sowie von LiH und MgH_2-Anwendungen von der organischen Synthese und als H_2-Speicher, Angew. Chem. 97 (1985), S.253-264, 116 Lit.) wird auf diese vielseitigen Moeglichkeiten eingegangen; einzelne Mechanismenvorschlaege werden unter Anwendung einer D-Markierung diskutiert.

Korrektur an
NATUR-REAKTOREN IN OKLO

In dem Beitrag "Praehistorisches Endlager", Isotopenpraxis 21(1985)9, gelbe Seiten II, ueber die Naturreaktoren in Oklo (Gabun) wurde durch einen Uebertragungsfehler angegeben, dass dort 12 000 t U-235 gespalten worden waren. In der Atomwirtschaft/Atomtechnik 30(1985)2, KuU I wird eine Korrektur darueber mitgeteilt. Es sind nicht t sondern 12 000 kg (mindestens) an U-235, die gespalten wurden, was einem Jahresbedarf von 12 grossen heutigen LWRen entspricht (und nicht 12 000 LWRen).

Veröffentlichungen / Publications

ZENTRALINSTITUT FUER ISOTOPEN- UND STRAHLENFORSCHUNG DER ADW DER DDR, LEIPZIG

Nr. 91 Information- und Rechentechnik fuer die Kernforschung; Ausgewaehlte Vortraege der KDT-Fachtagung ueber die Anwendung der Rechentechnik in der Inf.- und Dokumentation fuer die Kernforschung - Juli 1984

Nr. 92 Untersuchungen von radikalischen Primaerprozessen mit Hilfe von ESR-Pulsradiolyse- und ESR-Laserphotolyse-Experimenten. Effekte der chemisch induzierten dynamischen Elektronenspinpolarisation - D. Beckert u.a., August 1984

Nr. 93 Zur chemischen Beeinflussung der Zerfallskonstante von Iod-125 - Ch. Eifrig, August 1984

Nr. 94 Zur Strahlenbehandlung von Arzneimitteln, Biochemikalien und Impfstoffen - Literaturstudie - W. Nordheim, Dezember 1984

Nr. 95 DOSKMF2 - ein Begriff zur computergestuetzten Bewertung von Strahlenfeldern in Gamma-Bestrahlungsanlagen - M. Remer, Dezember 1984

Nr. 96 Traceruntersuchungen katalytischer Reaktionen von Kohlenwasserstoffen - J. Dermietzel, Dezember 1984

Nr. 97 Third Working Meeting on Radiation Interaction, Part I - J. Schmidt u.a. December 1984

Nr. 98 desgl., Part II - J. Schmidt u.a. December 1984

Nr. 99 Beitraege zur Chemie der Lanthaniden und Transplutoniumelemente - H. Bruchertseifer u.a.

Isotopenvariationen im Wasserhaushalt des Menschen in bezug auf D - H. Huebner, December 1984

Nr. 100 Wissenschaftliches Kolloquium "35 Jahre DDR - 15 Jahre Zentralinstitut fuer Isotopen- und Strahlenforschung der ADW" - Oktober 1984

Nr. 101 Beitraege zur Laserionisations-Massenspektroskopie - H.-J. Dietze, S. Becker, April 1985

Nr. 102 Wissenschaftlicher Ergebnisbericht 1984 des ZfI - Mai 1985

Nr. 103 14th International Symposium on Autoradiography - edited C. Treutler, July 1985

Nr. 104 Haeufigkeitsverteilung von Molekuelionen des Sulfid-Graphit-Systems in Funken-Massenspektren - S. Becker, H.-J. Dietze

Untersuchungen zum Ionenaustauschverhalten von Rh III aus salzsauren und salpetersauren Loesungen - L. Voigt, G. Pfrepper

Radiometrische und spektrometrische Untersuchungen der Loesungsmittel- und Milieuabhaengigkeit in den Systemen Palladium/Dithizon und Kupfer/Dithizon - H. Wagler, Oktober 1985

Nr. 105 Ueber die Roentgenemission in Abhaengigkeit von der Anregungsart und der chemischen Umgebung - E.Hartmann u.a.

On the electron rearrangement accompanying multiple ionisation by the impact of heavy charged projectiles - E. Hartmann

Zur Anwendbarkeit von K_a/K_β-Roentgenlinien-Intensitaetsaenderungen - G. Brunner u.a., August 1985

Nr. 106 Geographische Variationen der isotopen Zusammensetzung leichter Elemente im tierischen, insbes. menschlichen Organismus - H. Huebner, Oktober 1985

Ausgewählte Beiträge / Recently published papers

ATOM (1985) 341

Vitrification of Sellafield wastes - B. Smith

The utility of underlying research - J. Williams

The management of radioactive wastes - a review; The technical progress achieved in recent years

A passage to Japan; How the papers reported a shipment of plutonium

ATOM (1985) 342

Social and aconomic impact of nuclear electricity in Britain - D. White

Nirex - the challenge - E. Hoyle

A bright future for HTRs?

Nuclear power in Canada

ATOM (1985) 343

Unterlying research into the oxidation of metals - H.V. Atkinson

REACTOR: a computer simulation for schools - D. Squires

Advanced traction batteries: problems of development - R.M. Dell

ATOM (1985) 344

Transport package drop testing at Winfrith - D. Davis

Wind energy research in the CEGB - D.J. Milborrow

World status of fast reactor development (Three authors from the IAEA review the situation worldwide)

Plans for packaging and transporting radioactive waste - J.R. Grover

ATOM (1985) 345

EEC moves on to experimental waste facilities (Report of a waste management conference on the results of five years'research in Europe)

Nuclear energy under pressure: public inquiries in Britain and energy hearings in Germany - K. Clement

The pressurised water reactor (Report of a conference held in Birmingham)

Nuclear power in the FRG

ATOMKERNENERGIE/KERNTECHNIK 46(1985)3

Schwerpunktthema: Beschleunigertechnik

Neuere Entwicklung in der Physik von Elektron-Positron-Speicherringen - E. Keil

Hochfrequenz-Quadrupol-Beschleunigerstrukturen - R.W. Mueller

Zur Zeitstruktur des Schwerionenstrahls der Beschleunigeranlage VICKSI - W. Pelzer, K. Ziegler

Ausbau der Beschleunigeranlage des Schweizerischen Instituts fuer Nuklearforschung (SIN) U.Schraber, T. Stammbach

Ein 2000-Ci-Tritiumgastarget fuer photonukleare Reaktionen bei hohen Energien - H.-J. Gasson, T. Reichelt

27 km Vakuumsystem fuer den grossen Elektronen-Positronen-Speicherring - O. Groeber

Die Karlsruher Ionenmikrostrahlanlage und ihre Anwendungen - D. Heck

Neuere Entwicklungen auf dem Gebiet der Magnetmesstechnik - J. Billan

Reaktorsicherheit:
Auswirkungen einer Dampfexplosion im Verlauf eines hypothetischen Kernschmelzunfalles bei einem Leichtwasserreaktor - M. Peehs et al.

Bestimmungen des schnellen Neutronenflusses mit Hilfe der Methode der spektralen Indizes - I. Goncalves

ATOMKERNENERGIE/KERNTECHNIK 46(1985)4

Sind alternative Containmentkonzepte sinnvoll zur Schadenseindaemmung nach Kernschmelzunfaellen - J. Czech et al.

Chemische Probleme auf der Gasseite von CO_2-gekuehlten Reaktoren - B.H.M. Billinge

Die Kugelfunktionsmethode als Loesungsansatz fuer die nichtlineare Boltzmann-Gleichung - F. Schuerrer

Finite-Elemente-Verfahren fuer Core-Berechnungen in R, Z-Geometrie - Y. Abushady

Nachweis von Instabilitaeten an einem Forschungsreaktor mittels Reaktorrauschanalyse - S.A. Ansari, S.K. Azayuddin

Aehnlichkeitsgesetze fuer homogene und Zweiphasen-Stroemungen unter Beruecksichtigung des Waermeuebergangs von einem beheizten Stab - S. Benedek

Dosimetrische Eigenschaften von Be-O-Detektoren - W. Kriegseis et al.

Neue Messergebnisse zur Freisetzung von Radionukliden beim Auslegungsstoerfall -Messleitungsbruch- an Druckwasserreaktoren - W. Morell et al.

Eine analytische Berechnung von Gamma-Dosen aus endlich ausgedehnten Abluftfahnen durch Taylor-Entwicklung - E. Maraziotis

Abschirmfaktoren von Panzern moderner Bauart gegen die Neutronenbomben und Kernspaltungswaffen - S. Sahin, A. Gsponer

ATOMWIRTSCHAFT/ATOMTECHNIK 30(1985)7

Leistungsfaehige deutsche Kerntechnik - K.H. Beckurts

Umweltfreundliche Kernenergie - K. Hillermeier

Zur Genehmigung der WA Wackersdorf - A. Dick

30 Jahre Kerntechnik - Entwicklung und politisches Umfeld - H.-H. Haunschild

Die deutsche Wiederaufarbeitungsanlage - G.H. Scheuten

25 Jahre Institut fuer Plasmaphysik - Teil 1: Grundliegende Untersuchungen zur Aufheizung und zum Einschluss von Plasmen - E. Fuenfer et al.

Teil 2: Toroidaler Plasmaeinschluss bei Fusionstemperaturen - G. Grieger

Verfuegbarkeit der Kernkraftwerke in der Welt im Jahre 1984 - A. Szeless, F. Oszuszky

IAEA-BULLETIN 27(1985)1

Radioactive materials transport:
Overview: Perceptions, perspectives, and proportions

Safety considerations for spent fuel transport: View from the USA - R.M. Jefferson

Irradiated fuel transport - update from France - B. Lenail

International regulation: Clarifying roles - R.B. Pope

Physically protecting nuclear material - the legal framework - Ha-Vinh Phuong

Transport worker exposures: How much in the UK? - Ken B. Shaw

Case history: Sinking of the Mont-Louis & nuclear safety - B. Augustin

Topical reports:
Nuclear medicine in developing countries - Mohamed Nofal

Radiopharmacy: New techniques spurring growth - H.V. Rutz

Dosimetry for quality control in radiation processing - J.W. Nam

Reassessing the nuclear liability regime - Ha-Vinh Phuong

Nuclear power risks: Report from WHO - M.J. Suess

BIOCHEMISTRY 24(1985)7

Aspartate β-decarboxylase from Alcaligenes faecalis: Carbon-13 kinetic isotope effect and deuterium exchange experiments - R.M. Rosenberg, M.H. O'Leary

Carbon isotope effect on carboxylation of ribulose biphosphate catalyzed by ribulosebiphosphate carboxylase from Rhodospirillum rubrum - C.A. Roeske, M.H. O'Leary

BIOCHEMISTRY 24(1985)8

High-resolution cross-polarization/magic angle spinning C-13 NMR ol intracytoplasmic membrane and light-harvesting bacteriochlorophyle-protein of photosynthetic bacteria - T. Nozawa et al.

BIOCHEMISTRY 24(1985)9

Secondary isotope effects and structure-reactivity correlations in the dopamine β-monooxygenase reaction: evidence for a chemical mechanism - S.M. Miller, J.P. Klimman

Photoaffinity labeling of chromatin using a tritiated (Azidoaryl) bis (acridinyl) spermidine photoprobe - P.E. Nielsen

BIOCHEMISTRY 24(1985)10

Multinuclear magnetic resonance studies of metal ion binding sites of phosphoglucomutase - G.I. Rhyn et al.

BIOFIZIKA (UdSSR) 30(1985)2

Proton spin relaxation in the cytochrome-C-water system - V.V. Ivanov et al.

Identification of signals in overlapping NMR spectra of proteins by the difference spin echo method - N.F. Sepetov, V.N. Bushuev

BIOORGANICHESKAYA KHIMIYA 11(1985)3

Radioreceptor assay: theoretical principles of the method - S.U. Zaitsev et al.

A facile method for preparing radioactive nitroxyl radicals - L.A. Neimon et al.

AMERIC. J. CLIN. NUTRITION 41(1985)4

The measurement of whole body water by water-O-18

dilution in newborn pigs – R.K. Whyte et al.

VOPROSY MEDIZINSKOI KHIMII 31(1985)2
A modified radiometric procedure for estimation of amine oxidase activity – U.A. Pekkel et al.

AMERIC. J. PHYSIOLOGY 248(1985)4
Recycling of amino acid label with prolonged isotope infusion: implications for kinetic studies – W.F. Schwenk et al.

Determination of gluconeogenesis in vivo with C-14 labelled substrates – J. Katz

CANADIAN J. BIOCHEM. AND CELL BIOLOGY 63(1985)3
The cross-linking of human Met-hemoglobin with [C-14] dimethyl adipimidate – R.J. Ferris, D.B. Smith

J. PHARMACY AND PHARMACOLOGY 37(1985)4
Disposition of radiolabelled suppositories in humans – M. Jay et al.

KOSMICHESKAYA BIOLOGIYA – AVIAKOSMICHESKAYA MEDIZINA 19(1985)2
Amino acid metabolism in men exposed to short-term head-down tilt combined with ultraviolet irradiation – A.S. Ushakov et al.

Sensitivity of mamallian tissues to prolonged effects of constant magnetic fields of high strength – G.V. Galaktionova et al.

EUROP. J. NUCL. MED. 9(1985)11
Effective dose equivalent from radiopharmaceuticals – L. Johannson et al.

Quantitative differences between the thyroid uptaka of I-123 and Tc-99m – J.A. Siegel et al.

ZHURNAL MIKROBIOLOGII, EPIDEMIOLOGII, IMMUNOBIOLOGII (1985)4
Use of solid-phase radioimmunoassay for the serological diagnosis of influenza – T.N. Kobyakova

BIOCHEM. PHARMACOLOGY 34(1985)8
A novel approach for the study of thromboxane A-2 and prostaglandin H-2 receptors using I-125-labelled ligands – P.V. Halushka et al.

LABORATORY AND CLIN. MED. 105(1985)5
Indium-111 toxicity in the human lymphocyte – E.B. Silberstein et al.

A simple method for measurement of intestinal calcium absorption in humans by double-isotope technique – M. Griessen et al.

BILD DER WISSENSCHAFT 22(1985)6
Der Nuklid-Generator: weniger Strahlen, weniger Risiko – C. Keller
(beschrieben werden der Th-, ^{81m}Kr-, ^{212}Bi-, ^{132}J-Generator und die Voraussetzungen fuer einen Einsatz der Generatoren in der Nuklearmedizin)

BILD DER WISSENSCHAFT 22(1985)7
Kernfussion: Heisser als die Sonne – U. Deker

CANADIAN J. CHEM. 63(1985)1
Secondary α-D kinetic isotope effects for the E2 reaction of the 2-phenylethyl halides with t-butoxide ion in t-butyl alcohol – P.J. Smith et al.

J. CHEM. SOC. II PHYS. ORG. CHEM. (1985)2
Primary and secondary β-D kinetic isotope effects in the 1,3-prototropic rearrangement of 1-methylindene using tertiary amines as catalysts – O. Matsson

J. MOLEC. SPECTROSCOPY 109(1985)2
Rotational spectra of mono-18O-substituted ozones – C. Chiu, E.A. Cohen

The far-IR-spectrum of deuterated H_2S : The ground state rotational constants of $D_2{}^{32}S$, $D_2{}^{34}S$ $HD^{32}S$ and $HD^{34}S$ – C. Camy-Peyret et al.

J. Chem. Phys. 82(1985)3
The kinetic isotope effects in the reaction of $O(^3P)$ with H_2, D_2, and HD – N. Presser, R.J. Gordan

Isotopic vibration – vibration energy exchange in HCl molecules at low temperatures – K. Shin

Dynamics of the $H+D_2 \longrightarrow HD+D$ – D.P. Gerrity, J.J. Valentini

ZHURNAL ORGAN. KHIMII 21(1985)1 (in Russian)
Quantitative description of aromatic compounds reactions with eletrophilic reagents. 5. Using the hyperbolic equation for the description of isotopic hydrogen exchange reactions of aromatic compounds and their analogs in acidic media – V.A. Koptyung, N.F. Salakhutdiuov

Kinetic isotopic effect during the linear orthoester radical fragmentation – E.V. Pastushenko et al.

CONTRIB. MINERALOGY PETROLOGY 88(1984)1/2
Evolution of the blueschist and greenschist facies rocks of Sifnos, Cychlader, Greece. A stable isotope study subduction – related metamorphism – A. Matthews, M. Schliestedt

Nd-Sr isotopic and geochemical systematics in Cambrian boninites and tholeiites from victoria, Australia – D.R. Nelson et al.

J. SEDIMENTORY PETROL. 55(1985)1
Isotopic variations in dolomite concretions from the Monterey formation, Calif. – J. Hennesy, L.P. Knauth

OKEANOLOGIYA 25(1985)1 (in Russian)
Formation conditions of veined phosphates from the Pacific seamounts based on isotopic data – V.P. Strizkov et al.

AMER. J. SCI. 285(1985)2
Geochronologic and isotopic constraints on Late Precambrian crustal evolution in the Eastern Desert of Egypt – R.J. Stern, C.E. Hedge

J. PHYSICS F METAL PHYSICS 15(1985)3
The effect of isotopic substitution on the critical temperature of the Niob-Hydrogen-System – T. Kuji, T.B. Flanagan

EARTH PLAN. SCI. LETTERS 72(1985)1
The Pu-240/Pu-239 ratio, a potential geochrometer – M. Koide et al.

ANGEW. CHEMIE 97(1985)2
Grundprinzipien der proteasekatalysierten Knuepfung der Peptidbindung – H.-D. Jakubke et al.

NATURWISSENSCHAFTEN 72(1985)1
H_2O content and K-Ar-Data of metamorphic rocks from the Urach Drilling Site – P. Zaun

NATURWISSENSCHAFTEN 72(1985)3
Osmotische Adaptation bei Algen – G.O. Kirst

Wie aussichtsreich ist die Suche nach γ-Pulsaren? – V. Schoenfelder

K/Ar-Datierungen zur Alterseinstufung lagerstaettenbildender Prozesse – O. Brockamp, M. Zuther

Spaltspuren in Zr – ihre Bedeutung fuer kontinentale Tiefbohrungen – P. Zaun, G.A. Wagner

Glucose utilization in the toad's brain during anesthesia and stimulation ascending reticular arousel system – A ^{14}C-Desoxyglucose study – Th. Finkenstaedt, J.P. Ewert

Isotopenpraxis 21 (1985) 12, S. 433 bis 438

Tracer Transport Modelling for Unsaturated Zone — Effect of Anion Exclusion and Immobile Water

S. M. Rao, S. K. Jain

(Isotope Division, Bhabha Atomic Research Centre Trombay, Bombay)[1])

The injected tracer method for the evaluation of mass flow of moisture in the unsaturated zone is based on the 'piston-flow' model. The assumptions inherent in this concept are that the dynamic behaviour of the tracer is similar to the moisture in transport and that there is no 'immobile' component of soil moisture. Our studies indicate that the above assumptions can be verified by suitable mathematical simulation of the experimental data. The multi-cell mixing model and the advection-diffusion model have been used to simulate the laboratory and field tracer profiles. The faster movement of the $[Co(CN)_6]^{3-}$ *ion in clayey soils as compared to tritium is attributed to the anion exclusion of the trivalent anion. The paper also discusses the simulation of tracer-profiles using the modified advection-diffusion model taking into account the mobile and immobile components of soil moisture.*

Zur Bestimmung des Massenflusses der Feuchte in der ungesättigten Bodenzone wird bei der Injektions-Tracermethode das „piston-flow" Modell verwendet. Dabei wird angenommen, daß sich der Tracer dynamisch ähnlich der Feuchte verhält und daß keine immobile Komponente in der Bodenfeuchte vorhanden ist. Unsere Untersuchungen konnten dies unter Anwendung einer geeigneten mathematischen Simulierung der experimentellen Daten bestätigen. Es wurde das Multi-Zell-Mischungsmodell und das Advektions-Diffusionsmodell verwendet; sowohl bei Laborversuchen als auch bei Feldversuchen. Die schnellere Beweglichkeit des $[Co(CN)_6]^{3-}$*-Ions in tonigen Böden verglichen mit Tritium wurde dem Anionenausschluß des dreiwertigen Anions zugeordnet. Außerdem wurde ein modifiziertes Advektions-Diffusionsmodell diskutiert, das mobile und immobile Komponenten der Bodenfeuchte mit berücksichtigt.*

Keywords

cobalt 60; cobalt complexes; flow models; soils; tracer techniques; tritium

1. Introduction

The rate of infiltration of water through the unsaturated zone in the soil is of interest for setting up a regional water balance. The measurement enables an estimation of the fraction of rainfall that goes to recharge the ground water reservoir. There are a number of direct and indirect methods to measure infiltration rates. Direct methods using lysimeters or infiltrometers disturb the natural conditions of flow and are often considered non-representative. The indirect method of measuring changes in the soil moisture profiles in the unsaturated zone, using either the classical gravimetric method or the neutron probe method, needs to be coupled with the computation of evapotranspiration with the attendant inaccuracies.

The tritium tracer method developed nearly seventeen years ago [1] provides an elegant and direct approach to infiltration studies. In this method, the soil moisture in a horizontal layer (either at the surface or at a depth) is tagged with the radiotracer. Tritiated water (HTO), being one of the isotopic species of the water molecule is most commonly employed as the tracer. Infiltration caused by rainfall or irrigation would then push the tagged layer downwards like a moving piston. The new position of the tagged layer is then located by a suitable radiation detection procedure.

If the tagged layer has moved a distance d from its initial position and m is the average moisture content between the initial and final positions of the tagged layer, it follows from the piston flow anology that a quantity $m \times d$ of water has infiltrated into the soil after tagging.

Besides the displacement of the tagged layer by soil moisture transport, there would be an inevitable spread of the tracer around its position of maximum concentration, mainly due to molecular diffusion.

It would be intersting at this stage to discuss a question often raised by many users. Would it not be more realistic from the point of view of the piston flow model to determine m for sufficient depth before injection and use the pre-injection value of $\overline{m}$ to calculate the infiltration $\overline{m} \times d$? Obviously this is the amount that has been pushed down as the tagged layer moved downwards. If so, it would imply that there would be no difference in the value of $\overline{m}$ before and after displacement since d is the same in either mode of calculation. This can happen only under the following conditions:

1. The soil in the zone of tracer displacement is at its so called field capacity during the entire period of investigation
2. Any addition of water results in a fast discharge restoring the soil to its field capacity.

In view of the above, we may have to conclude that the piston-flow model is strictly not true when the soil moisture is below the field capacity at the place of injection.

Widely used tracer for soil moisture transport studies has been tritium. The main disadvantage with tritium is that the tracer is not amenable to 'in situ' detection. Recently gamma emitting Cobalt-60 as $K_3[Co(CN)_6]$ has been applied under field conditions after detailed laboratory investigations [2]. When injected around an access tube, ^{60}Co provides the advantage of 'in situ' and non-disturbing

[1]) Address: Bombay 400085, India

detection. The tracer displacement can be directly followed simply by γ-logging the access tube. The same tube can be used for recording soil moisture distribution using a neutron probe.

In clayey soils, however, the hexacyanocobaltate appears to move ahead of tritium. This necessitated further investigations on the transport characteristics of $[Co(CN)_6]^{3-}$. The faster movement of cobaltate ion with respect to HTO can either be attributed to anion exclusion of the trivalent $[Co(CN)_6]^{3-}$ or to the retardation of tritiated water due to ^3H exchange with the 'immobile' component of soil moisture.

In this paper with the help of suitable mathematical models we answer the following specific questions.

1. Is the piston-flow assumption valid?
2. How to explain the faster movement of $[Co(CN)_6]^{3-}$ with respect to HTO.

2. Mathematical model simulation of tracer profiles

Both the multi-cell model and the advection-diffusion model have been used to simulate the tritium and $[Co(CN)_6]^{3-}$ profiles obtained in field and laboratory studies.

a) *Multicell model*

The analytic solution of the multibox model having n boxes in series for a point injection of the tracer into the first box is given by

$$\frac{C_n}{C_0} = \frac{\left(\frac{Q}{V}t\right)^{n-1} e^{-(Q/V)t}}{(n-1)!} \tag{1}$$

where Q is the infiltration rate
C_n is the concentration of the tracer in the nth box
V is the moisture volume in each box

An iterative solution of the multi box model has been derived [3] and is given by

$$\frac{C_n^N}{C_0} = \frac{(N+n-2)!}{(N-1)!\,(n-1)!} \quad \frac{I^{n-1}V^N}{(I+V)^{N+n-1}} \tag{2}$$

where C_n^N is the conc. of the tracer in the nth box after N iterations of recharge pulses of value I.

b) *Advection-diffusion*

One dimensional advection diffusion equation describing solute transport is of the form

$$\frac{\partial C}{\partial t} = D\frac{\partial^2 C}{\partial x^2} - v\frac{\partial C}{\partial x} \tag{3}$$

where C is the solute concentration at distance x and time t, v is the pore-water velocity and D is the disperion coefficient describing the effects of molecular diffusion and hydrodynamic disperion. D is given by

$$D = \tau D_0 + v\alpha \tag{4}$$

where D_0 is the diffusion doefficient of the solute in free solution, τ is the tortuosity of the granular medium, and α is the dispersivity of the medium. The analytical solution of equation (1) for a point injection of an amount A of the tracer is

$$C(x,t) = \frac{A}{\sqrt{4\pi Dt}} e^{-(x-vt)^2/4DT} \tag{5}$$

The above mathematical models have been applied to simulate the tracer profiles obtained for HTO and $[Co(CN)_6]^{3-}$ in laboratory column studies using clayey soils and in one of the field studies at Wagoda in Maharashtra [3]. Tracer field profiles obtained for HTO by others [4] have also been simulated. Tracer profiles are shown in Fig. 1–4. The various fitting parameters etc. used are shown in Tab. 1–3.

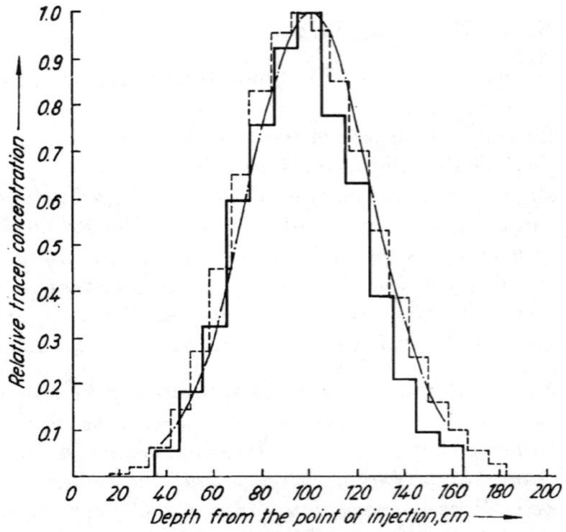

Fig. 1. Tritium Profile of Parbatpura (Sabarmati Basin)
——— Field; ------ Multibox model; —·—· Advection-diffusion model

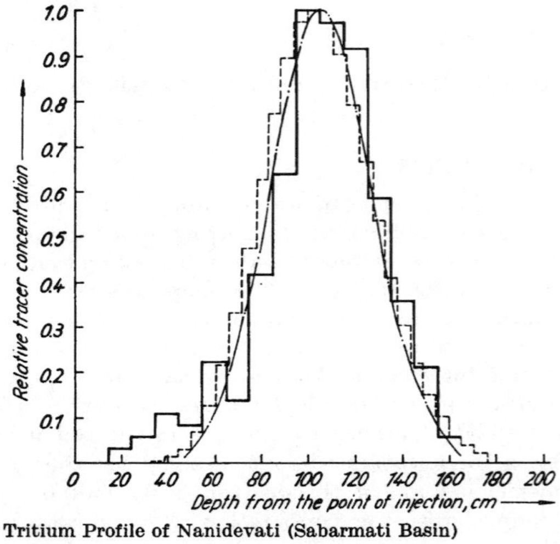

Fig. 2. Tritium Profile of Nanidevati (Sabarmati Basin)
——— Field; ------ Multibox model; —·—· Advection-diffusion model

Tab. 1. Field Experiments at Sabarmati basin by Bhandari, Gupta and Sharma PRL, Ahmedabad (India) [4]

Tracer	Model simulated	Average moisture content by vol	Fitting parameters	Recharge using model	Recharge from piston flow
HTO at Parbatpura	Multibox (Analytical)	0·213	$\frac{Qt}{V} = 11{\cdot}5$	20·4 cm	21·3 cm
	Advection-diffusion		$D = 3{\cdot}0$ cm²/d	21·3 cm	21·3 cm
HTO at Nani-Devati	Multibox (Analytical)	0·078	$\frac{Qt}{V} = 30$	8·05 cm	8·19 cm
	Advection diffusion		$D = 1{\cdot}5$ cm²/d	8·19	8·19

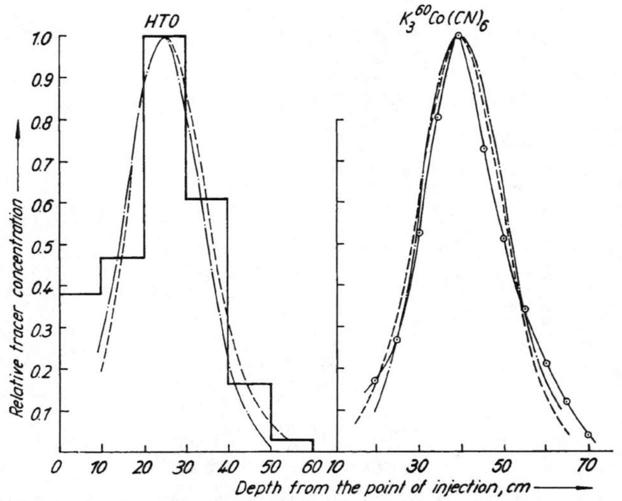

Fig. 3. Laboratory column experiment with clayey soil in unsaturated condition
——— Experimental; - - - - - - Multibox model; ·—·—· Advection-d ffusion model

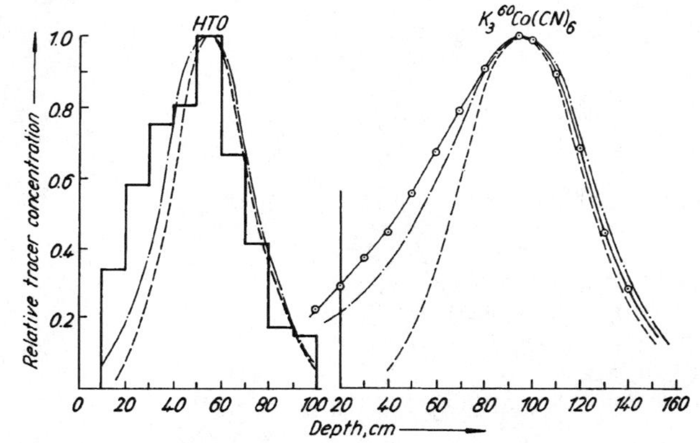

Fig. 4. Tracer profiles of wagoda
——— Field; - - - - - Multibox model; ·—·—· Advection-diffusion model with immobile water fraction

It can be seen that in all the cases the tracer transport depicts piston flow. One important observation is that the cobalt peak is always ahead of the tritium peak (in clayey soil).

3. Explanation of the faster movement of $[Co(CN)_6]^{3-}$ compared to HTO

In the above model simulation of the tracer profiles, two observations can be made:

1) Whereas tracer profiles obtained with tritiated water in clayey soils could be simulated using the total measured moisture content the tracer profiles obtained with $K_3[^{60}Co(CN)_6]$ using the same moisture values showed higher recharge. This means that for the same recharge value to be obtained with $[Co(CN)_6]^{3-}$ the moisture content will have to be reduced. Let us discuss tracer field profiles of Wagoda (Tab. 3).
In case of HTO the recharge calculated from the piston flow is $55 \times 0{\cdot}24 = 13{\cdot}2$ cm.
In case of $K_3[^{60}Co(CN)_6]$, the recharge calculated from the piston flow is $95 \times 0{\cdot}24 = 22{\cdot}8$ cm.

If we assume (to be proved later) that tritiated water transport is more representative of moisture transport and that the true recharge value is 13·0 cm, then it can be shown that the effective moisture content participating in the transport of

$[Co(CN)_6]^{3-}$ is 0·14

That is to say that out of a total measured moisture content of 0·24, only 0·14 or less than 60% of moisture participated in the cobaltate ion migration and 40% was excluded.

If on the other hand, we assume that $[Co(CN)_6]^{3-}$ was more representative of moisture transport and that HTO was being retarded due to exchange with immobile water then the piston flow concept demands that the effective moisture content in tritium transport be 0·4 much higher than the measured total moisture content of 0·24. This is not possible. Hence the possible explanation for the faster

Tab. 2. Laboratory Experiment in Unsaturated Condition with Clayey Soil

Tracer	Model simulated	Average moisture content by vol	Fitting parameters	Recharge using model	Recharge from piston flow
HTO	Multibox (Analytical)	0·4	$\frac{Qt}{V} = 6{\cdot}5$	9·3 cm	10·0 cm
	Advection-diffusion		$D = 2{\cdot}0$ (cm²/day)	10·0	10·0
$K_3[Co(CN)_6]$	Multibox (Analytical)	0·4	$\frac{Qt}{V} = 15{\cdot}5$	15·5 cm	16·0 cm
	Advection-diffusion		$D = 2{\cdot}5$ cm²/d	16·0	16·0

Tab. 3. Field Experiments at Wagoda (Maharashtra)

Tracer	Model simulated	Average moisture content by vol	Fitting parameters	Recharge using model	Recharge from piston flow
HTO	Multibox (Analytical)	0·24	$\frac{Qt}{V} = 10{\cdot}5$	12·6 cm	13·2 cm
	Advection-diffusion		$D = 1{\cdot}0$ cm²/d	13·2	13·2
$K_3[^{60}Co(CN)_6]$	Multibox (Analytical)	0·24	$\frac{Qt}{V} = 12{\cdot}5$	21·75	22·8
	Advection-diffusion		$D = 2{\cdot}0$ cm²/d	22·8	22·8

movement of $[Co(CN)_6]^{3-}$ is anion exclusion [5]. Anion exclusion is a phenomenon closely associated with the repulsion of negatively charged anions from the vicinity of the surfaces of clay particles which are also negatively charged. This zone of exclusion of anions reduces the tracer flow cross-section compared to water flow cross-section and the tracer ions move at the centre of the pore with a velocity higher than the mean pore water velocity.

2) It is observed that the effective diffusion coefficient for $[Co(CN)_6]^{3-}$ is generally larger than that of HTO. Also comparatively a longer tail has been observed in the case of $[Co(CN)_6]^{3-}$.

4. Advection-diffusion model — immobile water component

For non interacting tracers, the following modification of the advection-diffusion model has been proposed [6] to take into account the possible presence of immobile water.

$$\Theta_m \frac{\partial C_m}{\partial t} + \Theta_{im} \frac{\partial C_{im}}{\partial t} = \Theta_{im} D \frac{\partial^2 C_m}{\partial x^2} - v \Theta_m \frac{\partial C_m}{\partial x} \quad (6)$$

$$\Theta_{im} \frac{\partial C_{im}}{\partial t} = K(C_m - C_{im}) \quad (7)$$

where Θ_m and Θ_{im} are the volumetric fractions of the soil filled with mobile and immobile water respectively. C_m and C_{im} are the corresponding tracer concentrations at time t, v is the mean pore water velocity and K is a mass transfer coefficient (T^{-1}).

In the above equation, advection-diffusive transport is confined to the mobile phase while the diffusion controlled transfer of tracer into and out of the immobile zone is assumed to be proportional to the concentration difference between the two zones.

It is not easy to obtain an analytical solution for the above equations. Numerical solutions have been obtained in a computer using the method of finite differences.

Fig. 5—6 show for various values of K, the tracer concentration profiles in the mobile and immobile zones for the hypothetical case of a point injection on the top of a very long soil column. It is assumed that $F = \frac{\Theta_m}{\Theta_m + \Theta_{im}} = 0.5$. It can be seen that profiles of C_m and C_{im} are considerably different for lower values of K and as K increases the profiles become similar and they are identical at high values of K. Since the tracer has to diffuse into and out of the immobile zone, the tracer profile in the immobile zone lags behind the profile in the mobile zone.

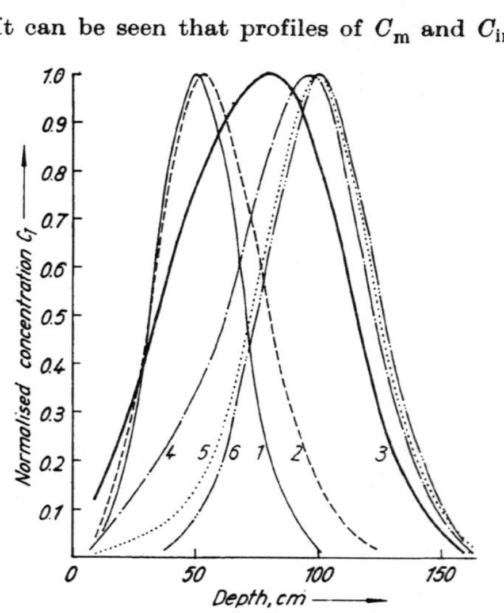

Fig. 5. Advection-diffusion-immobile component model: tracer profiles in the mobile phase

$F = 0.5$; $D = 1.5\ cm^2/d$; $v = 0.79$ cm/d 1 — $K = 0.1$; 2 — $K = 0.01$; 3 — $K = 0.002$; 4 — $K = 0.001$; 5 — $K = 0.0005$; 6 — $K = 0$

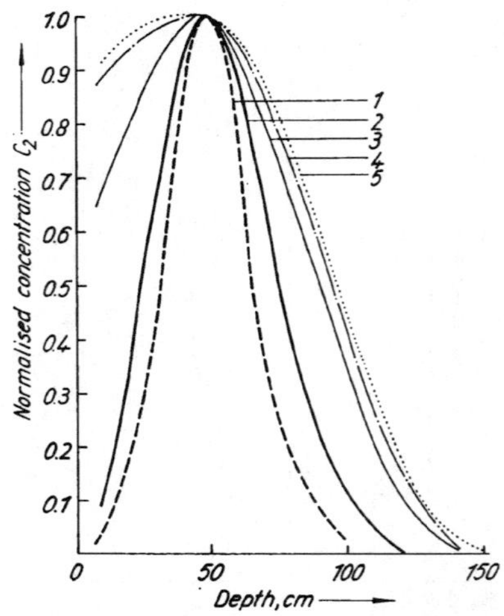

Fig. 6. Advection-diffusion-immobile component model: tracer profiles in the immobile phase

$v = 0.79$ cm/d; $F = 0.5$; $D = 1.5\ cm^2/d$ 1 — $K = 0.1$; 2 — $K = 0.01$; 3 — $K = 0.002$; 4 — $K = 0.001$; 5 — $K = 0.0005$

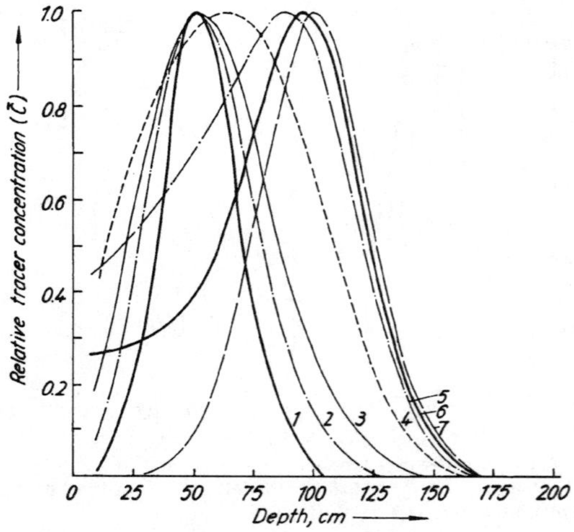

Fig. 7. Advection-diffusion model with immobile water fraction Weighted mean concentration profiles

$D = 1.5\ cm^2/d$; $F = 0.5$; $v = 0.79$ cm/d; 1 — $K = 0.1$; 2 — $K = 0.01$; 3 — $K = 0.005$; 4 — $K = 0.002$; 5 — $K = 0.001$; 6 — $K = 0.005$; 7 — $K = 0$

The weighted mean concentration of the tracer at any point in the transport medium is

$$\bar{C} = FC_m + (1 - F)\, C_{im} \quad (8)$$

The weighted mean concentration profiles (as obtained in column and field experiments) for various values of K are shown in Fig. 7. It can be seen that for $K = 0$ and for large values of K, the profiles are near gaussian whereas for intermediate values of K, they have a longer tail. When $K = 0$, the tracer transport is governed by the simple advection-diffusion equation (3). At large values of K when there is complete and instant mixing of the tracer in the two zones, the situation is again similar to Eq. (3), but the velocity of flow is reduced due to the increase in the effective cross section of tracer transport.

Modelling of the Wagoda data

The advection-diffusion model with immobile water component has been applied to simulate the Wagoda profiles and the results are given in Fig. 4. The best fits have been obtained with the following parameters.

Tab. 4.

Parameter	HTO	$Co(CN)_6]^{3-}$
Diffusion coefficient D	1·0 cm²/d	3·0 cm²/d
Mobile water fraction $F = \frac{\Theta_m}{\Theta_m + \Theta_{im}}$	0·95	0·5
Exchange coeff. K	0·0002/d	0·0003/d
Mean velocity	0·4 cm/d	0·75 cm/d

It is interesting to discuss the various differences between the two tracers.

The mobile water function $F = 0{\cdot}95$ for HTO indicates that there was no significant immobile water component in the Waghoda unsaturated zone. In the case of cobaltate ion, $F = 0{\cdot}5$ confirms the earlier interpretation that the tracer underwent anion exclusion and the excluded volume of moisture behaved as an 'apparant' immobile component.

The large difference in the diffusion coefficients was also observed in the earlier simulations. This can only be qualitatively interpreted as follows:

Due to anion exclusion, the bulk of the cobaltate ions moves at the centre of the pore space where the velocity is highest. However, there is some diffusion of the tracer ($K = 0{\cdot}0002$) into and out of the slower moving water close to the particle surfaces. This diffusion causes a larger apparent dispersion, necessitating a higher dispersion coefficient for curve fitting in the case of the cobaltate ions.

It can be seen from Fig. 4 that the advection-diffusion model with exchange between the tracer and the excluded zones provided a better fitting for the tailing part of the cobalt profile.

The above interpretation leads us to the conclusion that the faster movement of the cobaltate ion is due to anion exclusion, and not because of tritium retardation due to exchange with immobile water.

The above concepts of piston flow and anion exclusion have been further verified in controlled experiments, as detailed below.

5. Laboratory column experiment under saturation condition:

A set of experiments has been carried out using clayey soil under saturated condition (unlike the unsaturated data given in Tab. 2). The experiments were designed to obtain the tracer breakthrough curves. In this case ^{58}Co has been used in place of ^{60}Co in $K_3[Co(CN)_6]$ as this has a short half life of about 65 days which is more suitable for laboratory experiments compared to longer lived ^{60}Co.

A 50 cm long and 3 cm diameter glass column was filled with about 270 g of soil to a height of 37·5 cm. A regulated flow rate of 6·8 ml/h was maintained through the column under saturated conditions. After obtaining equilibrium flow conditions, both HTO (0·5 μCi) and $[^{58}Co(CN)_6]^{3-}$ (5 μCi) were injected together (0·5 ml) at the top of the soil solumn. Effluent water samples were collected at regular intervals and their tracer concentrations were determined. ^{58}Co was determined by NaI scintillation counting and tritium concentration was measured by liquid scintillation counting. The breakthrough curves are presented in Fig. 8.

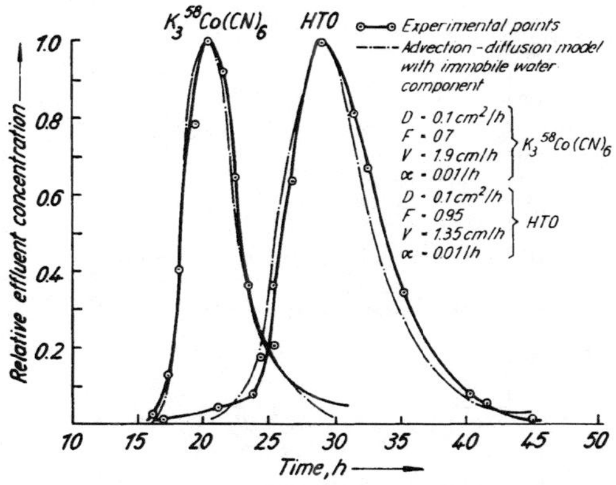

Fig. 8. Breakthrough curves for a clayey soil-laboratory column experiment
Correction: Instead of α read K

From the breakthrough curves, it can be seen that the tritium peak appeared nearly 29 h after injection, close to the calculated transit time of about 28 h. On the other hand, the cobalt peak appeared after 20 h, much earlier than expected.

The breakthrough curves obtained have been simulated using the advection diffusion model with immobile water component. Fig. 8 shows the profiles simulated and compares them with the experimental profiles. The best fits have been obtained with the following parameters.

Tab. 5.

Parameter	HTO	$[Co(CN_6)]^{3-}$
D	0·1 cm²/h (2·4 cm²/d)	0·1 cm²/h (2·4 cm²/d)
F	0·95	0·7
K	0·01 h⁻¹	0·01 h⁻¹
v	1·35 cm/h	1·9 cm/h

The K value had to be raised to 0·01 h^{-1} to fit the tail portion. At all lower values including $K = 0$, it was seen that good fit was obtained for most part of the c-t curve except at the tail. This is because of high velocities of flow. The bulk of tracer body moved fast so that any diffusion into and out of any immobile or slow moving anion-excluded zone affected only the tail.

6. Dry column experiment

Another laboratory column experiment was carried out to confirm the effect of anion exclusion on the cobaltate-ion. In this experiment, a continuous injection of tracer solution containing HTO and $[^{58}Co(CN)_6]^{3-}$ was made at the top of a 15 cm long dry clayey soil column. Effluent concentrations were determined and are presented in Fig. 9. It can be seen that the effluent front was highly enriched in cobaltate ion and the concentration slowly approached the injection value. This can be explained as follows:

As the water front moves down the column, the water wetting the soil is prodominantly devoid of the trivalent cobaltate ion due to anion exclusion and hence the percolating water is correspondingly enriched in cobaltate. As the injection continues, mixing due to disperion controls the cobaltate concentration in the percolating water which finally reaches the injection value. On the other hand, the non-ionic tritiated water does not display any such enrichment.

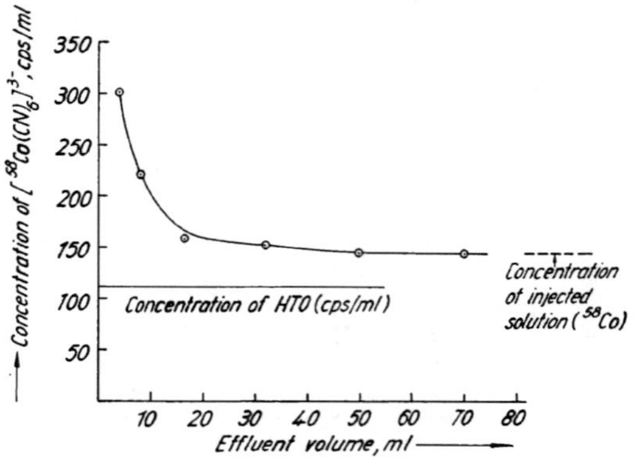

Fig. 9. Dry clay soil column experiment

7. Discussion and conclusion

The high F value of 0·95 in the field and laboratory experiments with clayey soils shows that there was no significant amount of immobile water in the unsaturated zone, the flow velocities are so small that there is complete mixing of an ideal, non-ionic tracer like $^1H^3HO$ over the moisture flow cross-section. This enables the tracer to fully represent the piston-flow mass transfer of soil moisture. The other significant observation is that anion exclusion is responsible for the faster movement of the trivalent cobaltate. Model-based corrections will be needed to estimate mass flow in the unsaturated zone when anionic tracers are used. Similarly studies on ion migration in the soils should include estimation of the effect of anion exclusion.

From the above the following conclusions may be drawn:

1) Piston flow concept of moisture movement in homogenous soils is largely valid.
2) Both the multicell and advection-diffusion models are well suited for application to solute migration in the unsaturated zone.
3) The cobaltate ion undergoes anion exclusion in clayey soils
4) There is no evidence of any significant presence of 'immobile' water causing retardation of HTO due to the exchange of tritium between the mobile and immobile zones.

Acknowledgements

Our thanks are due to our colleagues *A. R. Nair*, *R. Aranha* and *S. V. Navada* for their help in carrying out the studies and for helpful discussion.

Received March 14, 1985

References

[1] *U. Zimmermann, K. O. Munnich, N. Roether, W. Kreutz, K. Schubach, O. Siegel*, Science **152** (1966) 346

[2] *A. R. Nair, A. S. Pendharkar, S. V. Navada, S. M. Rao*, Groundwater recharge studies in Maharashtra – Development of isotope techniques and field experience. In: Isotope Hydrology 1978, Vienna: IAEA 1979, p. 803–826

[3] *S. M. Rao*, Use of radioactive tracers in studies on infiltration through unsaturated zone. Tracer Methods in Isotope Hydrology, TEC DOC-291, Vienna: IAEA 1983, p. 47–66

[4] *N. Bhandari, S. K. Gupta, P. Sharma*, Prem Sagar, *V. Ayachit, B. I. Desai*, Hydrogeological investigations in Sabarmati basin using radioisotope methods. Report No. HYD-82-01 Ahmedabad, PRL. 1982

[5] *G. H. Bolt, F. A. M. De Haan*, Anion Exclusion in Soils. In: Soil Chemistry, B. Physico Chemical models Ed. *G. H. Bolt*. Elsevier Scientific Publishing Company 1979, p. 233–257

[6] *J. P. Gandet, H. Jegat, G. Vachand, P. J. Wierenga*, Soil Science Soc. Am. J. **41** (1977) 665

Isotopenpraxis 21 (1985) 12, S. 438

Buchbesprechung

Die Welt der kleinsten Teilchen – Vorstoß zur Struktur der Materie

O. Höfling/P. Walaschek. Reinbek: Rowohlt Verlag 1984 511 S., 210 Abb., 68 Tab.; geb. DM 42,–

Wer sich heute ein Bild über die Struktur der Materie machen will, muß in spezielle Gebiete der Physik eindringen, die einerseits äußerst kompliziert sind und die zum anderen einer ständigen und vor allem sehr schnellen Erweiterung unterliegen. Über diese neuen Wissenskomplexe sind zusammenfassende Darstellungen kaum vorhanden, und die entsprechenden wissenschaftlichen Originalpublikationen werden naturgemäß von einem sehr großen Kreis potentieller Interessenten oft so abstrakt empfunden, daß sie für diesen Zweck als ungeeignet beiseite gelegt werden.

Um den Leser an den heutigen Stand der Teilchenphysik heranzuführen, handeln die Autoren äußerst geschickt das gesamte Gebiet der Struktur der Materie – von *Demokrit* über *Planck* bis zu den letzten Entdeckungen und den neuesten Theorien mit Kapiteln über die Atomhüllen und die Quantentheorie, die Atomkerne und die Subnuklearphysik – ab, um dann der Teilchenphysik, wie sie seit 1974 verstanden wird, einen besonders breiten Platz einzuräumen. Ein Kapitel unter dem Titel: „Ein Blick in die Zukunft" und einige weitere Lektüreempfehlungen schließen das insgesamt sehr gelungene Werk ab.

Die Autoren haben mit Erfolg versucht, die Begriffswelt der modernen Teilchenphysik in Worte zu fassen, die sie auch dem naturwissenschaftlich anspruchsvoller interessierten „Laien" verständlich machen. Sie verzichten ihrer Überzeugung nach weitestgehend auf mathematische Formulierungen. Wo es möglich war, ist an deren Stelle auf Modelle und Analogien zurückgegriffen worden, die zwar wertvolle Vorstellungshilfen sind, aber keinesfalls der Realität entsprechen.

Mit diesem Buch werden dem Leser sehr viele informative Fakten angeboten, womit das Ziel der Verfasser, einen Beitrag zur Überbrückung der immer größer werdenden Kluft zwischen dem Denken der Physiker und dem ihrer Zeitgenossen zu leisten, fraglos erreicht werden dürfte. Infolge seiner großen Informationsdichte, verbunden mit einem ausführlichen Sach- und Namenregister, sollte es aber auch als Nachschlagewerk gute Dienste verrichten. Es spricht nicht nur den Naturwissenschaftler und Ingenieur an, sondern auch Studenten und nicht zuletzt den Laien, der sich mehr als nur allgemein mit naturwissenschaftlichen Problemen auseinandersetzt.

J. Holzhey

Isotopenpraxis 21 (1985) 12, S. 439 bis 441

Untersuchungen zur Isotopenfraktionierung des Stickstoffs in den Systemen Ammonium/Ammoniak und Nitrid/Stickstoff unter geologischen Aspekten

H.-M. Nitzsche, G. Stiehl

(Zentralinstitut für Isotopen- und Strahlenforschung der Akademie der Wissenschaften der DDR)[1]

Für die Interpretation der beobachteten Zusammenhänge zwischen Stickstoffmenge und deren isotopischer Zusammensetzung in Metamorphiten und magmatischen Gesteinen ist die Kenntnis der Trennfaktoren für relevante Isotopenaustauschprozesse von Bedeutung. Auf quantenmechanischer Grundlage wurden deshalb von verschiedenen Autoren interessierende Isotopieeffekte berechnet. Ziel unserer Untersuchungen war es, einige dieser berechneten Gleichgewichte experimentell zu bestimmen. Im System NH_4^+/NH_3 *reichert sich das Nuklid* ^{14}N *in der Gasphase an. Im untersuchten Temperaturbereich bestätigen die experimentellen Ergebnisse im wesentlichen die theoretischen Berechnungen. In den Systemen* CrN/N_2 *bzw.* Ca_3N_2/N_2 *reichert sich das Nuklid* ^{15}N *in der Gasphase an. Dabei wurden selbst bei hohen Temperaturen* (>1000 K) *noch beachtliche Isotopieeffekte gemessen.*

For the interpretation of connections between the amount of nitrogen and its isotopic composition in metamorphites and igneous rocks the knowledge of the separation factors of some interesting isotopic exchange processes is essential. On the basis of quantum mechanics several authors calculated interesting isotope effects. The aim of our investigation was to determine experimentally some of these calculated equilibrium constants. In the system NH_4^+/NH_3 *the nuclide* ^{14}N *is enriched in the gaseous phase. In the systems* CrN/N_2 *and* Ca_3N_2/N_2 *the nuclide* ^{15}N *is enriched in the gaseous phase.*

Keywords

geochemistry; isotope effects; isotope ratio; nitrogen 14; nitrogen 15; stable isotopes

1. Einführung

Die beobachteten Schwankungen in der isotopischen Zusammensetzung verschiedener Elemente in geologischen Materialien sind das Ergebnis eines komplizierten Zusammenwirkens einzelner Isotopieeffekte, die sich auf physikalisch-chemische und physikalische Prozesse zurückführen lassen. Geologische Prozesse laufen meist in solchen Zeiträumen und unter physikalisch-chemischen Bedingungen ab, die auch eine Einstellung von Gleichgewichten bei Isotopenaustauschvorgängen gestatten. Für die Erklärung der unterschiedlichen isotopischen Zusammensetzung und die daraus abzuleitenden Schlußfolgerungen ist die Kenntnis der Größe von Isotopieeffekten (elementaren Trennfaktoren α) der in Betracht kommenden Austauschprozesse erforderlich.

Mit Hilfe der von *Urey* [1] eingeführten Methode der statistischen Mechanik ist es möglich, die Gleichgewichtsverteilung der stabilen Isotope zwischen zwei Phasen oder Verbindungen bei beliebigen Temperaturen zu berechnen. Voraussetzung dafür ist jedoch die Kenntnis verschiedener Moleküleigenschaften, insbesondere die der Grundschwingungsfrequenzen.

Derartige Berechnungen liegen für verschiedene stickstoffhaltige Verbindungen vor [2—4]. Für geologische Probleme ist das System $NH_4^+/NH_3/N_2$ von großer Bedeutung, da allgemein angenommen wird, daß der Stickstoff in Gesteinen und Mineralen hauptsächlich als Ammoniumstickstoff vorliegt.

Das Wirken von Austauschprozessen, an denen Nitride beteiligt sind, wurde von *Kung* und *Clayton* [3] für Meteoriten diskutiert. *Stiehl* [5] sowie *Stiehl* u. a. [6] diskutieren aufgrund neuerer experimenteller Untersuchungen an natürlichen Proben das Wirken derartiger Austauschreaktionen auch für die Lithosphäre.

Ziel der von uns durchgeführten Untersuchungen war es, die berechneten Fraktionierungseffekte für die Systeme Ammonium/Ammoniak und Nitrid/Stickstoff gemäß

$$^{14}NH_4^+ + {}^{15}NH_3 \rightleftharpoons {}^{15}NH_4^+ + {}^{14}NH_3 \quad (1)$$

$$^{15}N^{3-} + {}^{14}N_2 \rightleftharpoons {}^{14}N^{3-} + {}^{14}N^{15}N \quad (2)$$

experimentell zu bestimmen. Um die Untersuchungsbedingungen reproduzierbar zu gestalten, wurden die Versuche nicht mit geologischem Material, sondern mit reinen chemischen Substanzen durchgeführt. Die Temperaturen wurden dabei so gewählt, daß

- sie den angenommenen geologischen Bedingungen nahekommen
- der Isotopenaustausch bis zur Gleichgewichtseinstellung in einer überschaubaren Zeit stattfindet und
- die Substanzen, wie z. B. Ammoniumsulfat, noch stabil sind.

2. Untersuchungen am System Ammonium/Ammoniak

2.1. *Experimentelles*

Zur Bestimmung der elementaren Trennfaktoren

$$\alpha = \frac{([^{15}N]/[^{14}N])_{Verb.1}}{([^{14}N]/[^{14}N])_{Verb.2}}$$

für das Stickstoffisotopenaustauschgleichgewicht zwischen Ammoniumsulfat und Ammoniak werden ca. 60 mg

[1]) Anschrift: 7050 Leipzig, Permoserstr. 15

Ammoniumsulfat in eine Glasampulle eingebracht. Die evakuierten Ampullen werden mit 4,5 ml Ammoniak gefüllt, abgeschmolzen und für die jeweilige Versuchsdauer auf 250 bzw. 350 °C erwärmt. Nach dem raschen Abkühlen werden die Ampullen in einer Vakuumapparatur geöffnet, und das Ammoniak wird abgetrennt. Nach der Überführung des Ammoniaks in festes Ammoniumsulfat wird dieses mit Kupferoxid zu molekularem Stickstoff für die massenspektrometrische Isotopenanalyse umgesetzt [7]. Diese Messungen wurden an einem Massenspektrometer Varian MAT 86 mit Doppeleinlaßsystem durchgeführt. Die ermittelten δ^{15}N-Werte sind auf Luftstickstoff als Standard entsprechend

$$\delta N^{15}\,[^0/_{00}] = \left(\frac{[^{15}N/^{14}N]_{Probe} - [^{15}N/^{14}N]_{Standard}}{[^{15}N/^{14}N]_{Standard}}\right)\cdot 10^3$$

bezogen.

Zu Kontrollzwecken wurde auch der feste Rückstand untersucht. Die Standardabweichung der δ^{15}N-Werte beträgt $\pm 0{,}5\,^0/_{00}$.

2.2. Ergebnisse

In Tab. 1 sind die Ergebnisse der Austauschversuche zwischen Ammoniumsulfat und Ammoniak zusammengestellt. Das eingesetzte Ammoniumsulfat hatte einen δ^{15}N-Wert von $10{,}3\,^0/_{00}$, das verwendete Ammoniak von $1{,}9\,^0/_{00}$. Für die Einstellung des Isotopenaustauschgleichgewichts des Stickstoffs zwischen Ammonium und Ammoniak bei 250 °C sind etwa 100 Stunden erforderlich, während es sich bei 350 °C bereits nach ca. 75 Stunden eingestellt hat. Der Trennfaktor beträgt für 250 °C $\alpha = 1{,}0143 \pm 0{,}0012$ (Mittelwert aus sechs Einzelbestimmungen, $t > 100$ h) und für 350 °C $\alpha = 1{,}0128 \pm 0{,}0006$ (Mittelwert aus zehn Einzelbestimmungen, $t \geqq 75$ h).

In Tab. 2 sind die berechneten und experimentellen Ergebnisse zusammengefaßt.

Berücksichtigt man die unterschiedlichen Ergebnisse, die die verschiedenen theoretischen Berechnungsverfahren erbringen, so liegt die Abweichung der experimentellen Bestimmungen in vertretbaren Grenzen.

Es kann festgestellt werden, daß sich das betrachtete Gleichgewicht im untersuchten Temperaturbereich in einer aus geologischer Sicht kurzen Zeit einstellt. Die Größe der theoretisch berechneten Trennfaktoren konnte im wesentlichen experimentell bestätigt werden.

Tab. 1. Ergebnisse der Austauschversuche zwischen Ammoniumsulfat (AS) und Ammoniak (NH_3)

Versuchsdauer t h	Temperatur °C	$\delta^{15}N_{NH_3}$ $^0/_{00}$	$\delta^{15}N_{AS}$ $^0/_{00}$	α Fehler: $\pm 0{,}001$
17	250	0,0	10,7	1,011
17	250	0,0	10,7	1,011
25	250	0,0	10,7	1,011
25	250	0,0	10,6	1,011
49	250	0,0	10,7	1,011
49	250	0,0	10,7	1,011
65	250	−0,4	11,0	1,011
73	250	−1,4	11,1	1,013
89	250	−2,1	11,0	1,013
89	250	−2,4	11,1	1,014
122	250	−2,2	11,1	1,013
122	250	−3,4	11,3	1,015
242	250	−4,3	11,5	1,016
242	250	−2,5	11.2	1,014
482	250	−4,1	11,2	1,015
482	250	−2,2	11,0	1,013
75	350	−1,6	11,2	1,013
75	350	−2,4	11,4	1,014
218	350	−1,0	10,9	1,012
290	350	−1,3	11,0	1,012
290	350	−1,7	10,5	1,012
386	350	−1,6	11,1	1,013
386	350	−1,6	11,1	1,013
429	350	−2,0	11,1	1,013
429	350	−2,0	11,2	1,013
761	350	−1,5	11,2	1,013

Tab. 2. Berechnete und experimentell bestimmte elementare Trennfaktoren in Abhängigkeit von der Temperatur für das Isotopenaustauschgleichgewicht zwischen Ammonium und Ammoniak

Temperatur °C	Elementare Trennfaktoren α			
	Scalan [2]	*Hanschmann* [4]	*Kung und Clayton* [3]	*diese Arbeit*
127	1,0216	1,0204	1,022	–
227	–	1,0150	–	–
250	–	–	–	1.0143 ± 0,0012
327	1,0104	1,0115	–	–
350	–	–	–	1,0128 ± 0,0006
527	1,0077	1,0073	–	–
727	–	1,0051	1,0049	–

3. Untersuchungen am System Nitrid/Stickstoff

3.1. Experimentelles

Bei Vorversuchen hatte sich gezeigt, daß die Verfolgung des Austausches zwischen einem Nitrid und molekularem Stickstoff Schwierigkeiten bereitet, die ihre Ursachen in der Beständigkeit bzw. Unbeständigkeit der Nitride haben. Deshalb wurden die Nitridbildungsreaktionen bei Stickstoffüberschuß untersucht.

Dazu wurde das jeweilige Metall gemeinsam mit Stickstoff in einem Bombenrohr eingeschmolzen und erhitzt. Nach Ablauf der Reaktionszeit wurden die Bombenrohre bei Reaktionstemperatur geöffnet, und die Gasphase wurde von der festen Phase durch Abpumpen getrennt. Der nicht umgesetzte Stickstoff und das gebildete Nitrid wurden anschließend zur Durchführung von Isotopenanalysen aufbereitet.

Für die Untersuchungen wurden zwei Nitride mit möglichst unterschiedlichen Bindungsverhältnissen ausgewählt: Calciumnitrid, das zu den salzartigen Nitriden gezählt wird, und Chromiumnitrid, das zu den Nitriden mit kovalenten Bindungsverhältnissen gehört. Mit beiden Metallen wurden die Versuche in oben angegebener Weise durchgeführt. Dabei waren zum Erzielen ausreichender Reaktionsgeschwindigkeiten hohe Temperaturen erforderlich. Die Untersuchungen am Chromiumnitrid wurden deshalb bei 900 °C und am Calciumnitrid bei 950 °C durchgeführt.

Bei diesen Temperaturen hat das Chromiumnitrid bereits einen merklichen Zersetzungsdampfdruck, so daß zwischen dem gebildeten Nitrid und dem überschüssigen Stickstoff eine Gleichgewichtseinstellung erfolgen kann. Andererseits besteht die Gefahr, daß sich gebildetes Nitrid beim Abpumpen des Gases zersetzt. Es konnte von uns jedoch während der kurzen Zeit des Abpumpens keine Nitridzersetzung beobachtet werden.

3.2. Ergebnisse

Tab. 3. enthält eine Zusammenstellung der Ergebnisse. In der letzten Spalte ist der reziproke Trennfaktor $1/\alpha$ angegeben; das bedeutet, daß das schwere Nuklid ^{15}N in der Gasphase und das leichte Nuklid ^{14}N in der festen Phase angereichert wird. Damit befinden sich die experimentell

Tab. 3. Ergebnisse zur Isotopenfraktionierung bei Nitridbildungsreaktionen in überschüssiger Stickstoffatmosphäre

Nitrid	Versuchsdauer t h	Temperatur °C	$\delta^{15}N_{N_2}$ ‰	$\delta^{15}N_{Nitrid}$ ‰	$1/\alpha$ Fehler: ±0,001
CrN	96	900	7,0	−23,3	1,031
CrN	217	900	12,1	−15,8	1,028
CrN	217	900	11,9	−14,6	1,027
CrN	381	900	12,8	−14,2	1,027
CrN	381	900	10,9	−16,5	1,028
CrN	550	900	13,2	−13,6	1,027
CrN	550	900	14,2	−13,5	1,028
Ca_3N_2	16	950	15,1	−10,9	1,026
Ca_3N_2	26	950	−0,8	−12,9	1,012
Ca_3N_2	26	950	−1,4	−15,2	1,014
Ca_3N_2	43	950	−0,9	−13,5	1,013
Ca_3N_2	216	950	2,8	− 5,9	1,009
Ca_3N_2	216	950	2,0	− 7,1	1,009
Ca_3N_2	480	950	1,0	− 6,5	1,008
Ca_3N_2	480	950	1,3	− 6,8	1,008
Ca_3N_2	695	950	0,2	− 6,8	1,007
Ca_3N_2	695	950	0,4	− 6,8	1,007

ermittelten Isotopieeffekte bezüglich der Richtung in Übereinstimmung mit den theoretisch berechneten. *Kung* und *Clayton* [3] geben für das System Titaniumnitrid bzw. Aluminiumnitrid/Stickstoff bei 727 °C einen reziproken Trennfaktor von 1,0099 bzw. 1,0089 an. *Hanschmann* [4] errechnete für das System Bornitrid/Stickstoff ebenfalls bei 727 °C einen reziproken Trennfaktor von 1,0072.

Die experimentell ermittelten α-Werte für das System Ca_3N_2/N_2 liegen somit unter Berücksichtigung der Temperaturabhängigkeit von α in ähnlichen Größenordnungen wie die von *Kung* und *Clayton* [3] bzw. *Hanschmann* [4] für die oben genannten Systeme berechneten α-Werte. Unerwartet groß ist jedoch der reziproke Trennfaktor im System Chromiumnitrid/Stickstoff. Deshalb wurden die experimentellen Bedingungen besonders kritisch überprüft.

Die Vermutung, daß bei den hohen Temperaturen die Quarzreaktoren für Stickstoff durchlässig werden und ein Diffusionsisotopieeffekt auftritt, konnte nicht bestätigt werden. Untersuchungen, die mit unterschiedlichen Fülldrücken bzw. Reaktoroberflächen durchgeführt wurden, führten zu vergleichbaren Ergebnissen. Ebenso führten Versuche, das Austauschgleichgewicht zwischen dem Stickstoff, der aus Chromiumnitrid beim Erhitzen freigesetzt wird, und dem unzersetzt gebliebenen Nitrid zu bestimmen, zu Ergebnissen, wie sie bei der Untersuchung der Nitridbildungsreaktion erhalten wurden.

Die experimentellen Ergebnisse legen die Vermutung nahe, daß die Isotopieeffekte bei den salzartigen Nitriden kleiner sind als bei jenen mit kovalenten Bindungen. Die Richtigkeit dieser Aussage muß jedoch durch weitere Untersuchungen geprüft werden.

4. Schlußfolgerungen

Die experimentellen Ergebnisse zum Austausch der Stickstoffisotope zwischen Ammonium und Ammoniak kommen den berechneten Werten nahe. Gleichzeitig schließen sie jedoch nicht aus, daß bei höheren Temperaturen größere Abweichungen derart möglich sind, daß die theoretisch berechneten Werte niedriger liegen als die in der Natur tatsächlich auftretenden Isotopieeffekte. Dies ist dann von Bedeutung, wenn diese Temperaturabhängigkeit des Gleichgewichts als Kalibrierung Verwendung finden soll (z. B. bei den temperaturabhängigen Prozessen der Regionalmetamorphose).

Die Gleichgewichtseinstellzeiten für Temperaturen von 250 °C bzw. 350 °C lassen darauf schließen, daß bei Temperaturen, wie sie z. B. bei der Kontakt-, aber auch der Regionalmetamorphose vorkommen (≧400 °C), der Isotopenaustausch noch wesentlich rascher verläuft.

Damit wird — was bei einer Reihe von Untersuchungen stillschweigend vorausgesetzt wurde — wahrscheinlich gemacht, daß bei geologischen Prozessen, wie z. B. bei der Regionalmetamorphose, die thermodynamischen Isotopieeffekte dominieren.

Die Untersuchungen an Nitriden sind in zweierlei Hinsicht interessant. Erstens stützen sie die Vermutung, daß in der Lithosphäre ebenfalls ein Isotopenaustausch zwischen Stickstoff und Nitriden stattfinden kann. Die in den Gesteinen vorhandenen Schwermetalle können mit Stickstoff stabile Nitride bilden, wenn die entsprechenden Bildungsbedingungen gegeben sind. Auch das Silicium ist in der Lage, Nitride bzw. Oxynitride zu bilden.

Zweitens bestätigen die Ergebnisse, daß auch bei hohen Temperaturen (≦1000 °C) noch Isotopieeffekte des Stickstoffs von ≈ −10 ‰ auftreten können, wie sie z. B. an Basiten und Ultrabasiten in neueren Untersuchungen beobachtet worden sind (bisher unveröffentlichtes Material).

Es bedarf sicher noch weiterer gezielter Untersuchungen, um entscheiden zu können, unter welchen konkreten geologischen Bedingungen der Isotopenaustausch des Stickstoffs zwischen Ammonium und Ammoniak bzw. Stickstoff und unter welchen geologischen Bedingungen ein Austausch zwischen Nitrid und Stickstoff bzw. Ammoniak zustande kommt. Beim Prozeß der Regionalmetamorphose herrscht offensichtlich der Austausch im System Ammonium/Ammoniak bzw. Stickstoff vor, während in Gesteinen magmatischer Herkunft beide Austauschprozesse gleichermaßen wirken können.

Wir danken Frau Chem.-Ing. *U. Günther* und Frau Chem.-Ing. *R. Wollschläger* für die Durchführung der experimentellen Arbeiten sowie Herrn Dr. *D. Rohmer* und Frau *R. Ehrhardt* für die Durchführung der massenspektrometrischen Messungen.

Eingegangen am 26. 11. 1984

Literatur

[1] *H. C. Urey*, J. Chem. Soc. **136** (1947) 562
[2] *R. S. Scalan*, The isotopic composition, concentration and chemical state of the nitrogen in igneous rocks, Ph. D. Thesis, University of Arkansas, 1957
[3] *C. C. Kung, R. N. Clayton*, Earth Planet. Sci. Lett. **38** (1978) 421
[4] *G. Hanschmann*, ZfI-Mitteilungen Nr. 41 (1981) 19
[5] *G. Stiehl*, Untersuchungen der Isotopenvariationen des Stickstoffs in Erdgasen, organischen Sedimenten, metamorphen und magmatischen Gesteinen als Beitrag zur Klärung der Genese des Stickstoffs in Kohlenwasserstofflagerstätten, insbesondere im Rotliegenden der Altmark, Dissertation B, AdW der DDR, ZfI Leipzig, 1980
[6] *G. Stiehl, D. Haendel, K. Mühle*, ZfI-Mitteilungen Nr. 41 (1981) 187
[7] Autorenkollektiv, ZfI-Mitteilungen Nr. 37 (1981) 39

Kurze Originalmitteilungen

Isotopenpraxis 21 (1985) 12, S. 442 bis 443

Synthesen radioaktiv markierter Verbindungen. 52. Mitt.*) Darstellung von ^3H-markierten Monohydroxybenzoesäuren

G. Kaiser, H. R. Schütte

(Institut für Biochemie der Pflanzen der Akademie der Wissenschaften der DDR, Halle/Saale)[1])

Keywords

labelled compounds; organic compounds; synthesis; tritium

Für Transformationsuntersuchungen in Pflanzen und pflanzlichen Zellkulturen benötigten wir radioaktiv markierte o-, m- und p-Hydroxybenzoesäure. Die [^{14}COOH]-markierten Verbindungen lassen sich aus Kalium- bzw. Natriumphenolat mit $^{14}CO_2$ herstellen [2], sind aber für unseren Zweck nicht gut geeignet, da in Pflanzen Decarboxylierung auftreten kann. Wilzbach-markierte Präparate enthalten meist viele Nebenprodukte und erfordern daher eine aufwendige und verlustreiche Reinigung. So erhielten *Hartung* und *Gräser* [3] bei der Reinigung von *Wilzbach*-markierter Salicylsäure nur etwa 2% der spezifischen Radioaktivität des Rohpräparates im gereinigten Produkt. Wir versuchten deshalb, ein für einfache Phenole beschriebenes Verfahren des katalytischen Austausches mit Tritiumwasser [4] für unsere Zwecke zu modifizieren.

Experimentelles

Für die Austauschmarkierung werden 200 mg der jeweiligen gereinigten Hydroxybenzoesäure und 200 µg Triethylamin in ein 2-ml-Rundkölbchen mit langem engem Hals gegeben, welches anschließend sorgfältig mit Stickstoff gespült wird, um den beim katalytischen Austausch störenden Sauerstoff zu entfernen. Nach Zugabe von 400 µl Tritiumwasser (13,3 GBq/mmol = 20 Ci/ml) wird das Reaktionsgemisch unter Stickstoff in einem Chloroform/Trockeneis-Bad ausgefroren und der Hals des Kölbchens schnell zugeschmolzen. Die Ampulle wird in einem mit Sand und Glycerol gefüllten 100-ml-Rundkolben 72 h lang auf dem Wasserbad erhitzt. Nach langsamem Abkühlen und Ausfrieren des Inhaltes wird die Ampulle vorsichtig geöffnet.

Zur Reinigung wird das Reaktionsgemisch in einen 100-ml-Rundkolben überführt. Die Entfernung des überschüssigen Tritiumwassers erfolgte mit Ethanol und Benzen durch azeotrope Destillation, labiles Tritium wurde durch 5maliges Lösen in je 25 ml Methanol und Abdestillieren des Lösungsmittels entfernt. Bei allen Destillationen wurde das Destillat in einer Kühlfalle aufgefangen und zur Sicherheit zwei Waschflaschen mit Phenylethylamin nachgeschaltet. Den Rückstand lösten wir in 10 ml 10%igem Natriumbicarbonat und extrahierten 2 mal mit je 10 ml Ethylacetat, das anschließend mit 10 ml 10%igem Natriumcarbonat reextrahiert wurde. Auf diese Weise ließen sich die gefärbten Verunreinigungen fast vollständig aus der wässrigen Phase entfernen. Nach Ansäuern mit 10%iger Salzsäure wurde 4 mal mit Ethylacetat ausgeschüttelt. Die vereinigten Ethylacetatphasen extrahierten wir erneut 2 mal mit 20 ml 10%igem Natriumcarbonat. Schließlich wurde die markierte Säure nach Ansäuern mit Salzsäure durch 2malige Extraktion mit Ethylacetat wieder in die organische Phase überführt. Die vereinigten Ethylacetatextrakte wurden mit 10 ml 1%iger Salzsäure und 3 mal mit 20 ml dest. Wasser gewaschen und anschließend über eine (1 × 10)-cm-Säule mit wasserfreiem Natriumsulfat getrocknet. Nach Entfernen des Ethylacetates im Vakuum bei 50 °C lösten wir in Ether und entfernten diesen im Vakuum. Nach 2 Stunden im Vakuum bei 50 °C ist auch die aus dem Ethylacetat freigesetzte Essigsäure entfernt. Die als Rückstand erhaltenen festen Produkte erwiesen sich dünnschichtchromatographisch auf Silufol UV_{254} in 3 verschiedenen Laufmittelsystemen als rein (Tab. 1).

Tab. 1. R_F-Werte der Hydroxybenzoesäure nach DC auf Silufol UV_{254} in verschiedenen Laufmittelsystemen

		Rf Werte der Hydroxybenzoesäuren		
		o	m	p
1	Benzen/Methanol/Eisessig 96 8 4	0,52	0,33	0,33
2	Chloroform/Ether/Eisessig 40 9 1	0,86	0,42	0,48
3	Benzen/Dioxan/Eisessig 90 25 4	0,80	0,54	0,50

Tab. 2. Ergebnisse der katalytischen Austauschtritierung von Hydroxybenzoesäuren mittels Tritiumwasser (13,3 GBq/mmol)

Hydroxybenzoesäure	Ausbeute %	spezifische Radioaktivität GBq/mmol	Austauschrate [4]
o	91,5	1,48	22,2
m	94,0	13,17	99,0
p	72,5	3,00	22,5

Nach der gleichen Methode wurden auch die entsprechenden ^2H-markierten Monohydroxybenzoesäuren hergestellt. Durch NMR-spektroskopische Daten konnte nachgewiesen werden, daß der Austausch bevorzugt in den der Hydroxylgruppe benachbarten Positionen erfolgt. Das zeigte sich in einer Abnahme der Protonensignale dieser

*) 51. Mitt. vgl. [1]

[1]) Anschrift: DDR-4010 Halle/Saale, Weinberg 3

Positionen. Bei der radioaktiven Markierung wurden die in Tab. 2 zusammengestellten Ergebnisse erzielt.

Versuche mit verschieden stark markiertem Wasser zeigten, daß die Austauschrate von der spezifischen Radioaktivität des eingesetzten Tritiumwassers nahezu unabhängig ist.

Eingegangen am 3. 4. 1985

Literatur

[1] *A. Unverricht, H. R. Schütte, J. Hartung, H. Gräser*, Isotopenpraxis **21** (1985) 101.
[2] *Y. A. Muray, D. L. Williams*, "Organic Syntheses with Isotopes", Part I. New York/London: Interscience Publishers 1958, p. 365
[3] *J. Hartung, H. Gräser*, Pharmazie **37** (1982), 523
[4] *G. W. Kirby, L. Ogunkoya*, J. Chem. Soc. (London) **1965**, 6914

Aus der Praxis

Isotopenpraxis **21** (1985) 12, S. 443 bis 444

Multiplexer zur Erfassung von Impulsraten mittels Digitaloszilloskop

Keywords

computer calculations; data processing; laboratory equipment; multi-channel analyzers; radioactive materials; radiation detectors; tracer techniques

Zur Modellierung von Trennapparaten werden vielfach an diesen Verweilzeitmessungen mittels radioaktiver Tracer durchgeführt. Für diese Verweilzeitmessungen ist es erforderlich, an verschiedene Stellen des Apparates Detektoren zur Aufnahme von Impulsraten-Zeit-Kurven, worauf sich die Verweilzeitmessungen mittels radioaktiver Tracer reduzieren, anzubringen [1]. Ein dabei auftretendes Problem ist die Datenerfassung und -ausgabe. Als Beispiel soll der zur Modellierung von Ionenaustauschersäulen erstellte Versuchsaufbau dienen [2]: Um das Verweilzeitverhalten von Ionenaustauschersäulen zu ermitteln, erfolgte eine Messung der Konzentrationsveränderung in der flüssigen Phase in Form von Impulsratenveränderungen des eingesetzten Tracernuklids am Ein- und Ausgang der Säule, indem an diesen Stellen je ein Detektor (Sonde VA-J-968) angebracht worden war. Diese Sonden waren jeweils an einen Strahlenmeßplatz, bestehend aus Hochspannung, Verstärker, Analysator und Zählgerät, gekoppelt. Zur weiteren Datenerfassung wurden diese Strahlenmeßplätze über einen Multiplexer, dessen Konzipierung und Realisierung sich als Kopplungsglied notwendig machte, an einen Vielkanalanalysator Typ NTA-1024 geschlossen (Blockschema Abb. 1).

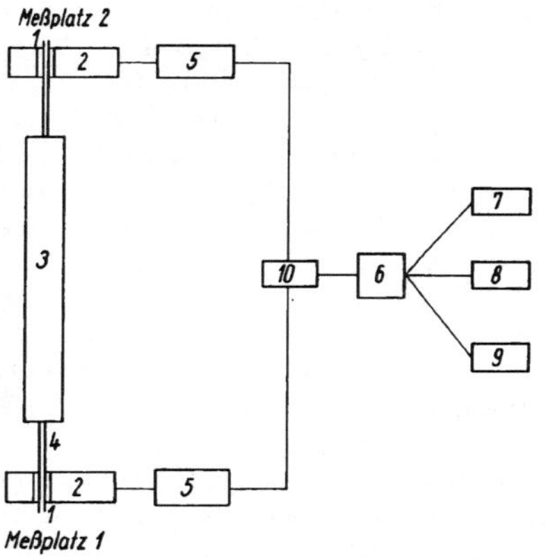

Abb. 1. Blockschema der Meßanordnung der Säulenanlage
1 – Na-J-Kristall; 2 – VA-J-968 Sonde; 3 – Ionenaustauschersäule; 4 – Flüssigkeitsleitung; 5 – Meßplatz 20026 m. Impulsdichtemesser; 6 – Vielkanalanalysator m. Bildschirm; 7 – Drucker; 8 – Lochstreifenstanzer EPLC 7191-01; 9 – Plotter; 10 – Zweikanalmultiplexer

Die Verwendung eines Vielkanalanalysators für Verweilzeitmessungen ermöglicht es erstens, eine beliebig kleine Meßzeit zu wählen, zweitens eine vielfältige Art der Datenausgabe, die entsprechend der Weiterbearbeitung auf einen Groß- oder auch Tischrechner erfolgen kann, und drittens, ein entscheidender Vorteil, eine kontinuierliche Messung während des gesamten Versuchsablaufs durchzuführen.

Die dem Ziel der automatischen Impulsabfrage- und Speicherung entsprechende Kopplung Strahlenmeßplatz-Vielkanalanalysator wurde möglich durch den Bau eines Zweikanalmultiplexers.

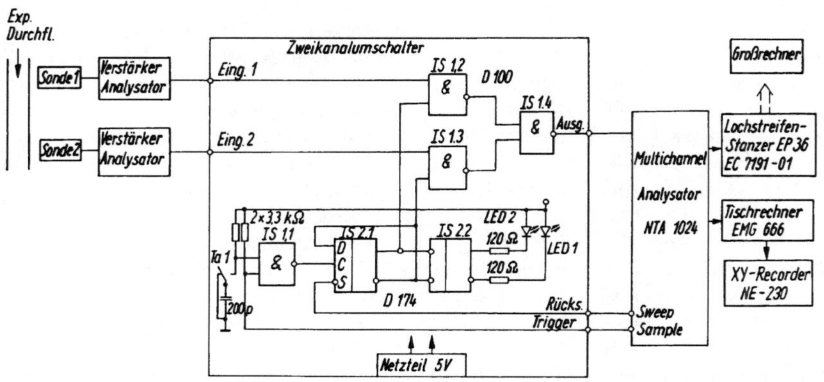

Abb.2. Gerätekonfiguration für 2-Sondenmessungen (Kanalumschalter detaillierter Stromflaufplan)

Schaltungsbeschreibung (Abb. 2)

Da sowohl der Ausgang der Verstärker/Analysatoren als auch der Dateneingang und die Steuerausgänge des Vielkanalanalysators TTL kompatibel sind, bietet sich eine Realisierung des Multiplexers mit TTL Bauelementen an. Die Gesamtschaltung zeigt Abb. 2. Die Impulse der beiden Meßplätze gelangen über die Tore $IS_{1.2}$ und $IS_{1.3}$ über $IS_{1.4}$ auf den gemeinsamen Ausgang. Durch das rückgekoppelte D-Flip-Flop $IS_{2.1}$ wird gesichert, daß jeweils immer nur ein Meßplatz auf den Ausgang durchgeschaltet ist. Welches Tor geöffnet ist, wird durch die beiden LED's über $IS_{2.2}$ angezeigt. Wird der Rücksetz- und der Triggereingang nicht beschaltet, können die Kanäle mit der Taste Ta 1 umgeschaltet werden. Im automatischen Betrieb wird die Kanalumschaltung vom Vielkanalanalysator vorgenommen. Dieser liefert in dem Mode „Digitaloszilloskop" bei jedem Speicherkanalwechsel am Ausgang „Sample" einen Impuls (high, 1 μs), der dann über $IS_{2.1}$ jeweils den anderen Meßplatz anschaltet.

Um eine feste Kanalzuordnung, z. B. Meßplatz 1 Kanal 1, 3, 5 ... 1023, Meßplatz 2 Kanal 2, 4, 6 ... 1024, zu erreichen, wird der „Sweep"-Ausgang (high während Digitalsweep) über den „Setz"-Eingang von $IS_{2.1}$ zur Synchronisation genutzt.

H. Fischer, D. Träber
Institut für Isotopen- und Strahlenforschung
der AdW der DDR, Leipzig
Eingegangen am 10. 4. 1985

Literatur

[1] *R. Löffel*, Atomkernenergie/Kerntechnik **33** (1979) 139

[2] *D. Mann*, Tracer research for modelling of ion-exchange equipment, 2nd Working Meeting "Radioisotope Application and Radiation Processing in Industry", 28th Sept. to 1st Oct. 1982, Leipzig, ZfI-Mitt. **69** (1983) p. 467 – 479

Isotopenpraxis **21** (1985) 12, S. 444 bis 446

Tagungsberichte

17. Internationales Symposium „Strahlenschutzphysik „vom 12. bis 15. März 1985 in Dresden

Das 17. Internationale Symposium „Strahlenschutzphysik" wurde vom Wissenschaftsbereich Strahlenschutzphysik der Sektion Physik der TU Dresden veranstaltet. Wissenschaftler aus vier sozialistischen Ländern (UdSSR, ČSSR, VR Ungarn und DDR) von insgesamt 18 Institutionen nahmen an diesem Symposium teil. In vier Themenkomplexe unterteilt wurden 26 Vorträge (Fortschrittsberichte, Forschungsberichte) gehalten, die neue Forschungsergebnisse und Entwicklungstendenzen auf dem Gebiet der Strahlenschutzphysik sowie Empfehlungen internationaler Organisationen behandelten.

Der erste Themenkomplex beschäftigte sich mit allgemeinen Problemen der Strahlenschutzphysik und der Meßtechnik im Strahlenschutz. *B. Dörschel* (DDR/Dresden) gab einen Überblick über die „Neueinschätzung der dosimetrischen Parameter von Hiroshima und Nagasaki und mögliche Konsequenzen für den Strahlenschutz". Anschließend diskutierte *K.-H. Weber* (DDR/Dresden) „Allgemeine Prinzipien der Strahlenschutzüberwachung — Kommentare zur ICRP-Empfehlung 35". *V. Bamblevskij* (UdSSR/Dubna) verlas einen Vortrag von *A. Shishkin* u. a. zum Thema „Die dosimetrische Überwachung mittels eines automatisierten Systems für die „F"-Anlage (DMASF)." *H.-U. Siebert* u. a. (DDR/Berlin) gaben einen „Überblick über Radioaktivitätsmessungen in Umgebungsmedien zur Einschätzung der Strahlenbelastung der Bevölkerung in der DDR". Am Ende des ersten Themenkomplexes stellten *O. Sarenio* und *R. Lehmann* (DDR/Berlin) „Passive Langzeitdosimetrie von Radon und seinen kurzlebigen Zerfallsprodukten basierend auf der elektrochemischen Ätzung von Polykarbonatdetektoren" vor.

Im zweiten Themenkomplex wurden Probleme der Lumineszenzdosimetrie behandelt. Die ersten beiden Vorträge waren Fragen der Bestimmung relativer Lichtkonversionsfaktoren von TL-Leuchtstoffen gewidmet. Nachdem *P. Schmidt* u. a. (DDR/Dresden) die „Experimentelle Basis zur Bestimmung von relativen Lichtkonversionsfaktoren von TL-Leuchtstoffen" erläutert hatten, stellten *K. Hübner* u. a. (DDR/Dresden) experimentell bestimmte „Relative Lichtkonversionsfaktoren für $CaSO_4$:Dy, CaF_2: Mn und LiF bei Bestrahlung mit Protonen im Energiebereich von 2 bis 9 MeV" vor. Anschließend referierte *M. Kubů* (ČSSR/Bratislava) über eine „Neue Methode zur Bestimmung don TL-Elektronentrapniveaus durch die Zwei-Temperatur-Behandlung". *P. Sahre* (DDR/Rossendorf) und *B.-U. Edelmann* (DDR/Berlin) berichteten über die „Simultane Auswertung der Dosis und der Daten einer Einzel-Bestrahlung durch Verwendung von $CaSO_4$: Dy-Umgebungsdosimetern". „Ein mikrorechnergesteuertes TLD-Auswertegerät in der Entwicklung — gegenwärtiger Stand und Richtungen" wurde von *H. Buschröder* und *M. Minor* (DDR/Berlin) vorgestellt. *K. Prokert* und *J. Fellinger* (DDR/Dresden) setzten sich mit „Problemen und Trends in der Präparation von mit seltenen Erden dotierten $CaSO_4$-Thermoluminophoren" auseinander. Abschließend gingen *J. Fellinger* u. a. (DDR/Dresden) auf den „Einfluß von unterschiedlichen Detektorparametern auf die Neutronenempfindlichkeit von OSL-Detektoren" ein.

Der dritte Themenkomplex beinhaltete Vorträge zur Problematik der Strahlenfelduntersuchung. *A. Hrabovcova* u. a. (ČSSR/Bratislava) beschäftigten sich in ihrem Vortrag mit „Beiträgen zur Kalibrierung von Personendosimetern in gemischten Neutronen-Gamma-Strahlungsfeldern". Danach behandelten *J. Hermanska* u. a. (ČSSR/Prag) die „Dosisäquivalentbestimmung mit Mehrkomponenten-Dosimetersystemen". *B. Horlbeck* und *D. Spitzbarth* (DDR/Berlin) diskutierten „Probleme der Bestimmung der für Kernkraftwerke typischen Gammastrahlung bis 8 MeV und Anwendung des Rechnerprogramms RADAK". Zur „Gammaspektrometrie bei hohen Zählraten" sprach *W. Sowa* (DDR/Berlin), während *M. Remer* (DDR/Leipzig) über „Ein Programm für die Berechnung von Strahlungsfeldern an Gammabestrahlungsanlagen" berichtete. *V. P. Bamblevskij* (UdSSR/Dubna) sprach anschließend über die „Bestimmung der Ausbeute und Winkelverteilung der Hadronen von mit ^{4}He und ^{12}C (3,65 GeV/

Nukleon) bestrahlten dicken Kupfertragets". *G. I. Britwich* u. a. (UdSSR/Serpuchov) beschäftigten sich in ihrem Vortrag mit „Neutronenfeldern für die Kalibrierung von Detektoren von Dosimetersystemen an kernphysikalischen Anlagen".

Der vierte Themenkomplex war neuen Tendenzen der Meßtechnik im Strahlenschutz gewidmet. *V. Alejnikov* u. a. (UdSSR/Dubna) stellten „Flußmesser und Dosimeter für Neutronen hoher Energie" vor. *A. G. Alekseev* u. a. (UdSSR/Serpuchov) berichteten anschließend über ein „Äquidosimeter ED-02 als Gerät für dosisäquivalente Messungen in gemischten Neutronen-Gamma-Strahlungsfeldern". Der Vortrag von *D. Zappe* u. a. (DDR/Dresden) behandelte „Einige Tendenzen auf dem Gebiet der Szintillationsdetektoren". *M. Danziger* u. a. (DDR/Dresden) referierten zu „Neutronennachweiseigenschaften von MOS-Sensoren". Die letzten drei Vorträge beschäftigten sich mit der Thematik der Elektretdetektoren. *B. Dörschel* und *G. Pretzsch* (DDR/Dresden) sprachen zu „Eigenschaften einer Elektretionisationskammer für die Routinedosimetrie in Photonenstrahlungsfeldern", während *G. Pretzsch* und *P. Wüstner* (DDR/Dresden) über die „Neutronenregistrierung mit Hilfe von Elektretimpulskammern" berichteten. Abschließend stellte *G. Pretzsch* (DDR/Dresden) „Auswerteprinzipien von Elektretdosimetern" vor.

Insgesamt kann eingeschätzt werden, daß das Vortragsprogramm einen Überblick über neueste internationale Entwicklungstendenzen auf dem Gebiet der Strahlenschutzphysik gab. Es wurden Empfehlungen internationaler Organisationen diskutiert und Anregungen für die weitere Forschungsarbeit gegeben. Darüber hinaus war der Austausch modernster Forschungsergebnisse auf dem Gebiet des Strahlenschutzes im Rahmen des RGW gewährleistet, was vor allem auf die internationale Beteiligung zurückzuführen ist. Das Symposium war durch ein hohes wissenschaftliches Niveau sowohl in bezug auf die Vorträge als auch auf die Diskussion gekennzeichnet. Verschiedene Vorträge des Symposiums wurden in einem Sonderheft (9/85) und nachfolgenden Heften der Zeitschrift „Kernenergie" veröffentlicht. Das 18. Internationale Symposium „Strahlenschutzphysik" findet vom **3.—7. Februar 1986** in Bad Schandau (in der Nähe von Dresden) statt.

G. Streubel

Eingegangen am 5. 6. 1985

7. Kolloquium „Technische Anwendung radioaktiver Isotope im Berg- und Hüttenwesen" am 13. und 14. März 1985 in Hetzdorf bei Freiberg

An diesem Kolloquium nahmen über 70 Fachleute aus der ČSSR, der Ungarischen VR, der VR Bulgarien, der VR Polen und der DDR teil. In 19 Vorträgen wurden vor allem Ergebnisse aus Arbeitsrichtungen der Isotopentechnik dargelegt und diskutiert, die gegenwärtig eine zunehmende Rolle insbesondere bei der Steigerung der Qualität zahlreicher Produkte des Industriezweiges und der Senkung von Verlusten spielen. Darüber hinaus wurde traditionsgemäß ein Übersichtsbericht über den Stand der Isotopentechnik im Berg- und Hüttenwesen eines der teilnehmenden Länder gegeben. Zwei Vorträge waren dem Strahlenschutz gewidmet, und in zwei weiteren Vorträgen ist der komplexe Einsatz der Isotopentechnik im Rahmen der BMSR-Technik zweier ausgewählter Betriebe vorgestellt worden.

Der Übersichtsbeitrag aus der VR Bulgarien demonstrierte eine überraschend große Breite in der Anwendung sowohl von Radiotracern zur Erforschung und Optimierung von Erzaufbereitungsvorgängen und metallurgischen Prozessen als auch kerntechnischer Geräte zur Kontrolle und Steuerung von Betriebsabläufen. Besonderes Augenmerk auf letztgenanntem Gebiet wird in der VR Bulgarien der schnellen Gehaltsbestimmung auf der Basis der isotopenangeregten Röntgenfluoreszenzanalyse geschenkt. Hierzu wurden drei für einen Serienbau vorgesehene Apparaturen vorgestellt, die sich hinsichtlich des gerätetechnischen Aufwandes und damit auch ihrer Leistungsfähigkeit deutlich unterscheiden und so eine für jeden Anwendungsfall ökonomische Auswahl zulassen.

Die Demonstration des komplexen Isotopeneinsatzes erfolgte an den Beispielen einer Zinnerzaufbereitung und eines Kalibetriebes. Die radiometrischen Bunkerfüllstandsmessungen, Skipentleerungskontrollen, kontinuierlichen Dichte-, Feuchte- und Gehaltsbestimmungen sind wesentliche Bestandteile der Meß- und Automatisierungstechnik in den Unter- und Übertagebereichen dieser Betriebe. In zum Teil mehr als 20jährigem Einsatz vor Ort haben die verwendeten Meßgeräte, die überwiegend vom VEB RFT Meßelektronik „Otto Schön" Dresden (MKD), zu einem großen Teil aber auch aus dem Forschungsinstitut für NE-Metalle Freiberg (FNE) unter Verwendung von Baugruppen von MKD stammen, ihre hohe Zuverlässigkeit und eindeutige Überlegenheit gegenüber nach anderen Prinzipien arbeitenden bewiesen.

Im Rahmen der Einzelbeiträge nahm der Einsatz von Radiotracern zur Aufstellung von Produktbilanzen, in der Korrosionsforschung an edelmetallbeschichteten Festanoden, zur Ermittlung des Stromwirkungsgrades und der Metallströmung in Al-Schmelzflußelektrolysezellen, zur Klärung von Erstarrungs- und Elementverteilungsvorgängen beim kontinuierlichen Guß von Kupfer, Aluminium und Stahl, für Untersuchungen über die Elementverteilung bei der Pfannenlegierung und in Forschungskomplexen zur Entwicklung und Optimierung hydrometallurgischer Verfahren für die Rückgewinnung von NE- und Edelmetallen aus Sekundärrohstoffen und industriellen Abwässern einen deutlich größeren Raum ein als bei den vorangegangenen Kolloquien. Das ist kein Zufall, sondern untermauert die Tatsache, daß mit Hilfe der Radiotracertechnik auf solch wichtigen Gebieten des Industriezweiges wie dem Recycling, dem kontinuierlichen Metallguß, der Energieverbrauchssenkung, der Qualitätsverbesserung, um nur einige zu nennen, wesentliche, unmittelbar praxiswirksame Beiträge geleistet werden. Aber es muß, wie in den Diskussionen vielfach zum Ausdruck gebracht wurde, auch an dieser Stelle nochmals betont werden, daß diese Möglichkeiten trotz vielfältigster Bemühungen der Isotopentechniker immer noch nur zu einem Teil genutzt werden.

Eine Zwischenstellung nehmen die Arbeiten zur Bestimmung der Verteilung und Konzentration von Bor mit Hilfe der neutroneninduzierten Autoradiografie, worüber in zwei Vorträgen berichtet wurde, insofern ein, als bei dieser Untersuchungsmethode der Tracer das im untersuchten Metall inaktiv vorliegende Bor selbst ist, das nur während der Zeitdauer der Neutronenbestrahlung im Kernreaktor die für den Bornachweis auf Festkörperspurdetektoren erforderlichen Alphateilchen emittiert.

Auf dem Gebiet der Anwendung umschlossener Strahlenquellen stand die isotopenangeregte Röntgenfluoreszenzanalyse mit Vorträgen über ein im FNE neu entwickeltes rechnerbestücktes Analysengerät und eine Reihe neuer Anwendungsbeispiele stark im Vordergrund.

Interessante und für den Teilnehmerkreis sehr lehrreiche Beiträge über den Strahlenschutz rundeten die Veranstaltung ab. Ausgehend von einem Überblick über den Stand des Einsatzes radiometrischer Strahleneinrichtungen im Berg- und Hüttenwesen wurden kurz, aber instruktiv die gesetzlichen Forderungen an deren Einsatz, die vom Betreiber zu erfüllenden Bedingungen für einen strahlenschutzmäßig sicheren Betrieb und einige Beispiele für aufgetretene außergewöhnliche Ereignisse, die beim Isotopeneinsatz im Berg- und Hüttenwesen der DDR ohne jeglichen Personenschaden verliefen, angeführt. Aus letzteren wurden wichtige Schlußfolgerungen über das ordnungsgemäße Verhalten beim Umgang mit radioaktiven Stoffen gezogen.

J. Holzhey

Eingegangen am 18. 6. 1985

VI. ^{15}N-Anwendungskursus der FAO/IAEA in Leipzig vom 13. 5.—8. 6. 1985

Die IAEA und die FAO hatten den „Interregional Training Course on the Use of ^{15}N in Soil Science, Plant Nutrition and Agricultural Biotechnology" zum sechsten Male hintereinander an die DDR vergeben. Hierzu waren Landwirtschaftswissenschaftler aus 14 Entwicklungsländern Afrikas, Asiens und Südamerikas zur Teilnahme ausgewählt worden. Im Auftrag des Staatlichen Amtes für Atomenergie und Strahlenschutz der DDR leitete wiederum Prof. *H. Faust* vom Zentralinstitut für Isotopen- und Strahlenforschung Leipzig der AdW der DDR den Kursus, unterstützt durch bewährte Fachleute dieses Instituts sowie durch offizielle Fachvertreter der IAEA und weitere Landwirtschaftsexperten der DDR.

Erklärtes Ziel dieser Kurse ist es u. a., die Entwicklungsländer in die Lage zu versetzen, ihre Pflanzenproduktion durch Optimierung der Stickstoffversorgung von Boden und Pflanze entscheidend zu verbessern. Die Teilnehmer trainieren vorwiegend die Methodik der ^{15}N-Markierung als eine wichtige Voraussetzung für diese Aufgabe.

Das straff organisierte Kursusprogramm umfaßte Vorlesungen, Seminare und Experimente zu folgenden Schwerpunkten:

— Grundlagen der ^{15}N-Tracertechnik (Isotopenverdünnungsanalyse, *Kjeldahl*-Methoden; Vorbereitung der Meßproben)

— ^{15}N-Isotopenanalyse (Massenspektrometrie; Emissionsspektrometrie)

— ^{15}N-Anwendung für Boden- und Pflanzenuntersuchungen (u. a. Harnstoffumwandlung im Boden; Stickstoff-Fixierung durch Pflanzen, Stickstoffverluste aus Boden und Düngemittel; Verteilung des Düngemittelstickstoffs auf verschiedene Pflanzenteile und Boden).

Anschließend hatte jeder Teilnehmer die Gelegenheit, spezifische landwirtschaftliche Probleme seines Landes zur Diskussion zu stellen („Country papers").

Zur Erleichterung des Beginns eigener ^{15}N-Untersuchungen nach der Rückkehr erhielt jeder Teilnehmer erstmals einen „^{15}N-Düngemittel-Kit". Dieser auf Anregung von IAEA-Vertretern in der DDR entwickelte und produzierte Kit enthält verschiedene ^{15}N-markierte Düngemittel sowie Eichproben und Umarbeitungschemikalien. Die Teilnehmer hatten weiterhin die Möglichkeit, am „NOI-6" zu trainieren, dem neuesten Typ der bewährten DDR-Serie der ^{15}N-Analysatoren auf emissionsspektroskopischer Basis.

Eine wertvolle Ergänzung des ^{15}N-Trainingsprogramms war die Besichtigung folgender Institute, wobei jeweils bestimmte Anwendungsschwerpunkte diskutiert wurden:

— Zentralinstitut für Isotopen- und Strahlenforschung Leipzig der AdW der DDR
— Institut für Biotechnologie Leipzig der AdW der DDR
— Institut für Genetik und Kulturpflanzenforschung Gatersleben der AdW der DDR
— Institut für Pflanzenernährung Jena der AdL der DDR
— Institut für Düngungsforschung Leipzig—Potsdam der AdL der DDR.

Die Teilnehmer und die anwesenden IAEA-Repräsentanten werteten abschließend den Kursus als sehr nützlich und erfolgreich.

P. Krumbiegel

Eingegangen am 1. 7. 1985

Isotopenpraxis 21 (1985) 12, S. 446 bis 456

Buchbesprechungen

Probleme der Geochronologie und der Isotopengeologie

Hrsg.: L. N. Ovčinnikov. Moskva: Nauka 1981. 196 S. (in Russ.); kart. M 11,—; R 2,20

Der vorliegende Sammelband enthält 15 Beiträge fast ausschließlich sowjetischer Autoren über geochronologische und isotopengeologische Untersuchungen an überwiegend präkambrischen Gesteinen der UdSSR sowie über damit zusammenhängende methodische Fragen. Das Hauptaugenmerk liegt auf physikalischen Altersbestimmungen an hochmetamorphen Gesteinskomplexen (z. B. der Halbinsel Kola), wobei mitunter eine erstaunliche Datenfülle auffällt. Daneben werden auch Untersuchungsergebnisse von Pegmatiten (K—Ar), Graniten (Rb—Sr, K—Ar), Erzbildungen (Pb-Isotope als Geneseindikatoren) sowie Basiten und Ultrabasiten (K—Ar) mitgeteilt. In den fünf mehr oder weniger stark methodisch orientierten Beiträgen geht es hauptsächlich um die Pb-Methoden (analytische Fragen der Pb-Abtrennung, Pb-Isotopendatierung von Zirkonen ohne chemische Abtrennung des Bleis oder anhand von Pb-Isotopenanalysen mittels Thermoionenemission, Probleme der Interpretation von Pb—Pb-Gesamtgesteinsisochronen); eine Publikation widmet sich den Möglichkeiten der Rekonstruktion unverfälschter K—Ar-Alter.

Wie der Titel des Sammelbandes hervorhebt, steht die Behandlung von Problemen der eingesetzten Datierungsmethoden im Vordergrund. Besonders deutlich demonstriert wird das am Beispiel der konventionellen K—Ar-Methode, mit der sich übrigens über die Hälfte aller Beiträge befaßt. Demgegenüber kommen Probleme der Rb—Sr-Geochronologie (2 Beiträge) relativ kurz weg. Selten, aber in einer Arbeit über geochronologische Untersuchungen (Rb—Sr, K—Ar) an vererzten Graniten recht gelungen praktiziert, werden voneinander unabhängige Altersbestimmungsmethoden kombiniert an ein und demselben geologischen Untersuchungsobjekt eingesetzt.

Bei der Zusammenstellung der Beiträge für diesen Sammelband wäre hinsichtlich der gegenwärtig häufig angewendeten radiogeochronologischen Methoden eine bessere Ausgewogenheit wünschenswert gewesen unter Einbeziehung modernerer Methoden wie der $^{40}Ar/^{39}Ar$- oder der Sm—Nd-Methode, die gerade in hoch- oder polymetamorphen (präkambrischen) Gebieten international mit Erfolg und in zunehmend größerem Umfang eingesetzt werden.

Den nicht so sachkundigen Leser könnten möglicherweise einige wenige (transskriptionsbedingte) Druckfehler bei Symbolen für chemische Elemente bzw. Nuklide verwirren (^{238}V statt ^{238}U, mehrfache Verwechslung von Rb und Pb und ihrer Nuklide), was aber den Wert des Sammelwerkes, der vor allem in dem publizierten neuen Datenmaterial und in der Behandlung von Problemen der konventionellen K—Ar-Datierung besteht, nicht beeinträchtigt.

Der vorliegende Sammelband ist sicher in erster Linie für Spezialisten gedacht, die sich mit Methoden und Problemen der physikalischen Altersbestimmung befassen. Darüber hinaus bietet er aber auch den regionalgeologisch interessierten Geowissenschaftlern neues Faktenmaterial über die geologische Entwicklungsgeschichte ausgewählter Gebiete der UdSSR.

U. Wand

Nuklearmedizinische Diagnostik maligner Tumoren Fortschritte der Onkologie Bd. 12

Hrsg.: H.-J. Altenbrunn. Berlin: Akademie-Verlag 1984. 321 S., zahlr. Abb. u. Tab.; kart. M 60,—

Die Diagnostik hat in der Onkologie bekanntlich einen besonders hohen Stellenwert. Da sich die nuklearmedizinischen Untersuchungsverfahren in den letzten drei Jahrzehnten einen festen Platz auf diesem Gebiet erworben haben, ist das Vorhaben, die Leistungsfähigkeit dieser Verfahren darzustellen, sehr zu begrüßen. Dankenswerterweise gelang es, eine Reihe von Spezialisten mit großer Erfahrung für bestimmte Untersuchungsverfahren zu gewinnen.

Nach einem kurzen einleitenden Überblick über wichtige Elemente der nuklearmedizinischen Tumordiagnostik (*H.-J. Altenbrunn*) werden von *G. Nauber* die technisch-physikalischen Grundlagen der nuklearmedizinischen Lokalisationsdiagnostik behandelt. Es wurde besonderer Wert auf eine allgemein verständliche Darstellung von Aufbau und Arbeitsweise der wichtigsten Geräte gelegt. *H.-J. Altenbrunn* geht in dem Kapitel über die Diagnostik von Schilddrüsenmalignomen speziell auf die Szintigrafie (mit Radioiodnukliden bzw. Pertechnetat) ein und widmet anschließend der Metastasendiagnostik einen besonderen Abschnitt. Nach der Besprechung der Diagnostik von primären und sekundären malignen Knochentumoren durch *J. Markwardt* und *A. Janisch* folgt ein Kapitel über die Diagnostik von Lebermalignomen speziell durch Szintigrafie (*H.-J. Altenbrunn*). Die nuklearmedizinische Diagnostik von Lungentumoren wird von *W.-G. Franke*, die von malignen Pankreastumoren von *J. Markwardt* dargestellt. *H. Deckart* gibt eine Übersicht über die Möglichkeiten, die durch die Nuklearmedizin bei der Diagnostik von Nebennierentumoren geschaffen wurden. In den folgenden Kapiteln widmen sich *W.-G. Franke* den Tumoren des Zentralnervensystems und *C. Mlatschkow* den malignen Erkrankungen der Lymphknoten. Nach der Besprechung der nuklearmedizinischen Diagnostik in der Ophthalmologie durch *K.-P. Lommatzsch* und *D. Hallermann* beschließt ein von *H.-J. Altenbrunn* vorgenommener Ausblick das Werk.

Jedes Kapitel ist in sich geschlossen und verfügt über eine eigene Literaturübersicht, die auch zahlreiche Arbeiten aus den 80er Jahren berücksichtigt. Bei jedem Verfahren werden Indikation, Kontraindikation und Strahlenbelastung genannt. Großer Wert wurde auch auf eine vergleichende Abschätzung der Leistungsfähigkeit nuklearmedizinischer und anderer möglicher Diagnosemethoden gelegt. Die sehr reichhaltige Ausstattung mit Abbildungen und Tabellen trägt wesentlich zum Verständnis der textlichen Ausführungen bei. Das Werk bietet dem onkologisch tätigen Arzt eine vorzügliche Einführung in die nuklearmedizinische Lokalisationsdiagnostik und gibt darüber hinaus jedem Interessierten die Möglichkeit, sich schnell über den modernen Stand auf diesem Gebiet zu informieren.

H. Herzmann

Rätselhafte Biographie der Erde

W. A. Drujanow. Leipzig: VEB Deutscher Verlag für Grundstoffindustrie 1984. 208 S., 17 Abb.; Leinen M 15,—

Hätte man vor einhundert Jahren eine Prognose gewagt zur Beurteilung des gegenwärtigen Kenntnisstandes über die Struktur und Dynamik unserer Erde, so wäre sie sicherlich zu positiv ausgefallen im Vergleich zu der Aussicht, auf dem Mond landen oder Raumsonden zu anderen Planeten schicken zu können. Wie dem auch sei, mit dem ansprechend aufgemachten Buch von *Drujanow* liegt ein allgemeinverständlicher Abriß zum aktuellen Kenntnisstand der Entwicklung unseres Planeten vor. An der Fülle der angebotenen Hypothesen erkennt der Leser, daß eigentlich noch vieles „offen“ ist. Dabei werden auch Hypothesen ausführlicher behandelt, die nicht Allgemeingut der Geowissenschaftler bei der Interpretation der Entwicklung unserer Erde, insbesondere ihres „Innenlebens“, darstellen. Das ist keinesfalls negativ zu werten; im Gegenteil bieten oft Gedanken, die sich nicht im Rahmen allgemein akzeptierter Denkmöglichkeiten und Modelle bewegen, den Anstoß, Neues zu entdecken. Hier sei zum Beispiel auf den Abschnitt „Unterirdische Gewitter“ und die angesprochene Möglichkeit hingewiesen, daß sich gasförmige Kohlenwasserstoffe (Methan u. a.) bei niedrigen Temperaturen auch infolge elektrischer Entladungen aus Erdöl bilden können.

Vom Autor werden alle Probleme angesprochen, die die geologischen Wissenschaften heute bewegen. Das vermittelt dem Leser einen ausgezeichneten Überblick und, sofern er besonders interessiert ist, auch eine weitergehende Beschäftigung mit einigen neuen Hypothesen.

Drujanow hat, insbesondere hinsichtlich neuer Hypothesen, vorwiegend sowjetische Autoren berücksichtigt. Bei der Bedeutung, die die geologischen Wissenschaften in der Sowjetunion und diese wiederum in den anderen Ländern haben, ist durchaus gesichert, daß sich die weltweit geführte Diskussion moderner geowissenschaftlicher

Probleme in der sowjetischen Fachliteratur widerspiegelt.

Das Buch stellt an den Leser keine besonderen Ansprüche hinsichtlich bestimmter Vorkenntnisse; es kann jedem an naturwissenschaftlichen Problemen Interessierten empfohlen werden.

H. Schütze

The Handbook of Environmental Chemistry Vol. 1, Part C: The Natural Environment and the Biogeochemical Cycles

O. Hutzinger (Ed.), Berlin — Heidelberg — New York — Tokyo: Springer-Verlag 1984. XIII + 220 S., 55 Abb., 9 Tab.: geb. DM 124.— ; US $ 46.30

Es ist dem Herausgeber dieser Handbuchserie — Professor *O. Hutzinger*, Bayreuth — zu danken, daß nicht nur neue Ergänzungsbände zu den einzelnen, abgegrenzten Volumina erscheinen werden, sondern daß auch eine Erweiterung der Volumina mit den Titeln Luft- und Wasserverschmutzung und Spurenanalytik in der Umwelt vorgesehen ist. Die Umweltchemie — weit verzweigt in den klassischen Wissensdisziplinen und von unterschiedlichsten Blickwinkeln aus betrachtet — bedarf einer Zusammenführung des bereits angehäuften Wissens. Was liegt näher, als diese Überschau in Form einer Handbuchreihe dem zunehmenden Interessentenkreis anzubieten, bei aller Problematik, mit der ein solches Unternehmen angegangen wird.

Der vorliegende Teilband C setzt den Themenkreis zur natürlichen Umwelt und den biogeochemischen Kreisläufen fort. Es werden 6 Kapitel mit den Überschriften Huminostoffe (strukturelle Aspekte), Organische Stoffe im Meer, Mariner Gelbstoff, Die Oberfläche des Ozeans, Atmosphärische Stickstoffverbindungen (Chemie, Nitrifikation, Denitrifikation und ihre Wechselbeziehungen) und schließlich Kohlendioxid — ein biogeochemisches Porträt — behandelt. Diese Aufzählung läßt erkennen, daß teils neue Aspekte zu den bisherigen hinzugetreten sind, aber auch bereits erkannte Lücken, die unbedingt einer Behandlung bedurften, ausgefüllt wurden, wie beispielsweise die Rolle der Stickstoffverbindungen in der Atmosphäre oder des Kohlendioxids. Dabei werden wie immer bei Kreislaufbetrachtungen in der Natur die Quellen, Senken und die Flüsse zu den Reservoiren herausgearbeitet, um so zu geschlossenen Darstellungen zu kommen.

Einige ergänzende Bemerkungen: Auf S. 109 wird das elektronisch angeregte Sauerstoffatom $O(^1D)$ mit seiner Anregungsenergie von 45,4 kcal/mol angeführt — es sollte mit einer 1 als Index gekennzeichnet werden, wie es allgemein in der Spektroskopie üblich ist. Außerdem ist die Hauptform seines Abbaues in der Troposphäre nicht die Löschreaktion mit O_2 und N_2, sondern in Gegenwart der Luftfeuchte die Bildung von OH-Radikalen, die auf diese Weise zu rund 90% entstehen. Im Inhaltsverzeichnis sind mehrmals die δ^{13}C-Werte nur als δ^{13} angegeben — dies kann zu Verwirrungen führen. Die δ^{13}C-Werte für Diamanten in der Abb. 6 auf S. 137 können durchaus den dort angezeigten Wertebereich von $-4 \ldots -7$ ‰ weit überschreiten; es sind auch Werte von $-26 \ldots -30$‰ bekannt.

Auch dieser Teilband zeichnet sich durch eine sehr hohe Auswertung der Literatur aus. Insgesamt wurden 1166 Literaturzitate gebracht, wobei 55% aller Literaturstellen allein aus den Jahren 1976—1984 stammen; den Hauptanteil bilden Arbeiten aus dem Jahre 1982 (8,3%). Es ist gelungen, Wissenschaftler anzusprechen, die sich einen Überblick auf diesem modernen Gebiet der angewandten Chemie verschaffen wollen; aber auch Spezialisten werden viele wertvolle Hinweise finden und sie für eigene Arbeiten verwenden können.

H. Hübner

Systemanalyse und Regelkreissynthese Eine einführende Darstellung auf der Grundlage der Übertragungsfunktion

E. D. Dickmann. Stuttgart: B. G. Teubner 1985. 272 S., 193 Abb. (Leitfäden der angew. Mathematik und Mechanik Bd. 60); geb. DM 58,—

Mit dem Begriff Systemanalyse wird gegenwärtig — leider nicht einheitlich — ein weites Spektrum von verbalen Beschreibungen komplexer Systeme bis zu strengen mathematischen Analysemethoden belegt. Der Autor des Buches versteht unter dem Begriff offensichtlich die in den letzten Jahrzehnten gut ausgearbeiteten Methoden der Analyse und Synthese von kybernetisch interessierenden Systemen auf der Basis der Regelungstheorie (bzw. deren modernem Äquivalent — der Systemtheorie). In diesem Sinne stellt das Buch eine solide Einführung in die Theorie des Gebiets dar unter besonderer Berücksichtigung der praktischen Anwendbarkeit unter Verwendung von Mikrorechnern. Dabei erfolgt die Darstellung zwar von einem modernen Standpunkt, zur ontogenetischen Veranschaulichung wird aber die phylogenetische Entwicklung des Wissenschaftsgebiet wirkungsvoll verschiedentlich demonstriert.

Das Buch enthält sechs Kapitel:

Nach einer anschaulichen Einführung in die historische Entwicklung der Regelungstechnik, die bis in das Altertum zurückreicht, folgt im zweiten Kapitel die Beschreibung dynamischer Systeme mittels Übertragungsfunktionen, dem eigentlichen Anliegen des Buches. Hier wird an überschaubaren Beispielen zunächst das Rüstzeug für die Konstruktion entsprechender (mathematischer) Modelle beschrieben, die Überführung der Modell-Differentialgleichungen in algebraische Gleichungen (mittels der *Laplace*-Transformation) dargestellt und die sich ergebende (analytische) System-Übertragungsfunktion im Komplexen diskutiert.

Das dritte Kapitel dient dem tieferen Eindringen in die Handhabung von Übertragungsfunktionen zusammengeschalteter Teilsysteme und charakterisiert Übertragungsfunktionen auf der Basis von (evtl. experimentell zu ermittelnden) Daten.

Im vierten Kapitel wird die „klassische" Regelungstheorie (nämlich die verschiedenen „Eingrößen"-Reglertypen) bis zu „modernen" Regelungskonzepten dargestellt.

Das fünfte Kapitel dient der Diskussion der Aspekte wie Regelungsstabilität gegen Störungen und Parameterschwankungen, wie sie den praktisch tätigen Ingenieur ständig interessieren.

Erst das sechste Kapitel vermittelt einen Ausblick auf Mehrgrößenregelungen.

Der Anhang enthält zwei kurze aber vollständige FORTRAN-Programme zur qualitativen Analyse bzw. Synthese von Reglern.

Das Literaturverzeichnis enthält im wesentlichen die weiterführenden Standardwerke des Fachgebiets der letzten Jahrzehnte.

Das Buch ist mathematisch vorgebildeten kybernetisch interessierten Studenten entsprechender Disziplinen zu empfehlen. Aber auch der Spezialist findet manche didaktische Anregung.

R. Straubel

Risks and Benefits of Energy Systems
Proc. of a Symposium. Jülich, 9–13 April 1984

Vienna: IAEA 1984. 671 S., 136 Abb., 158 Tab. (Proc. Series STI/PUB/668); kart. ÖS 1300.–

Im April 1984 fand in Jülich (BRD) ein internationales Symposium über Gefährdungen und Vorteile unterschiedlicher Energiesysteme statt. Diese Tagung wurde gemeinsam mit der UNEP (UN Environment Programme) und der WHO von der IAEA veranstaltet. Es nahmen 180 Wissenschaftler aus 36 Ländern und 11 internationalen Organisationen teil. Im vorliegenden Proceedings-Band sind 31 Vorträge, 8 Poster und 3 Panel-Diskussionen enthalten.

Die Sicherung des ständig steigenden Energiebedarfs geschieht in den einzelnen Ländern auf unterschiedliche Weise, wobei jedes Land vor ganz unterschiedlichen Folgen der Energiegewinnung steht. Daß eine gleichzeitige Analyse von Risiken für die Menschen und für die Umwelt einerseits und die ökonomischen Vorteile bestimmter Energiegewinnungen andererseits erforderlich ist, bedarf der Einbeziehung landesspezifischer Parameter unterschiedlichster Art. Auf diesem Symposium sollte – und wurde auch – eine Kommunikation hergestellt werden, die verallgemeinerungsfähig Aussagen zu diesem gegenwartsnahen Problem gestattet. Kosten/Nutzen-Analysen unter Herabsetzung der Gefährdungen sind die Zielfunktionen, die im einzelnen zu betrachten sind, wobei eine Trennung von industriell hochentwickelten Ländern und Entwicklungsländern vorgenommen werden muß. Daß neben einer vertieften systemanalytischen Betrachtungsweise noch bestimmte Einstellungen – Attitüden – eine nicht unwesentliche Rolle spielen, wurde in manchen Beiträgen deutlich gemacht.

Dieser Proceedings-Band dürfte für Energiewirtschaftler und Kernenergetiker eine sehr wertvolle Schrift sein, aus der nicht nur aktuell prognostizierte Energiesteigerungsraten bis zum Jahre 2030 zu entnehmen sind, sondern viele Hinweise gegeben werden, wie andere Länder die anstehenden energiewirtschaftlichen Forderungen bewältigen – alles unter dem Aspekt der Rentabilität bei geringsten Belastungsfolgen für die Natur.

H. Hübner

Nuclear Power Plant Instrumentation and Control: A Guidebook

Vienna: IAEA 1894. 368 S., 145 Abb., 17 Tab. (Technical Reports Series No. 239); kart. ÖS 780.–

Das vorliegende Handbuch für Instrumentierung, Steuerung und Kontrolle (sog. ,,Leittechnik") von Kernkraftwerken ist auf Leichtwasserreaktoren ausgerichtet. Zusätzlich sind auch Informationen über den Druck-Schwerwasserreaktor CANDU aufgenommen worden. Es ist nach den Angaben des Herausgebers für ,,Leittechnik"-Fachleute ausgearbeitet worden, die mit Planung und Beschaffung sowie mit der Vorbereitung der Inbetriebnahme und des Routinebetriebes eines KKW beauftragt sind.

Neben einem Überblick über die aktuellen Probleme auf dem Gebiet der Leittechnik von KKW werden direkte Hinweise für die einschlägigen Aufgaben im Rahmen eines KKW-Projektes gegeben.

Im ersten Teil des Handbuches werden alle Probleme der Leittechnik diskutiert, die in den verschiedenen Phasen eines KKW-Vorhabens auftreten – während der Planung des Projekts, der Beschaffung, der Prüfung, des Aufbaus, der Inbetriebnahme und des normalen Leistungsbetriebes. Diese Probleme betreffen hauptsächlich Organisationsfragen (Zeitplan, Verantwortlichkeiten u. a.), Genehmigungsfragen, Kaderfragen u. a. Die rechtzeitige Ausbildung der Kader, die Absicherung der regelmäßigen Auffrischung der Kenntnisse und die rechtzeitige Einrichtung eigener Ausbildungsplätze, wie z. B. eines Trainingssimulators in vollem Maßstab, spielen dabei eine wichtige Rolle. Es werden recht konkrete Empfehlungen gegeben, z. B. für die Kaderzusammensetzung in den verschiedenen Aufbau-Phasen des KKW oder für die anzuwendenden Meßmethoden, Angaben, die natürlich entsprechend den nationalen Besonderheiten modifiziert werden müssen.

Im zweiten Teil wird ein Überblick über die Probleme der Leittechnik von KKW gegeben, und es werden die aktuellen Konzepte dargestellt. Die verschiedenen existierenden Lösungen werden dabei allerdings nicht kritisch bewertet, sondern von den verschiedenen Herstellern aus Kanada, Frankreich, Schweden, der BRD und Japan in einem Anhang zum Handbuch nebeneinander dargestellt. Dabei wird sehr deutlich auf viele grundlegende Erfahrungen hingewiesen, die in einem modernen Projekt berücksichtigt werden sollten, z. B. auf Fragen der Zuverlässigkeit und Instandhaltung (Standardisierung in der Ausrüstung, Service-Freundlichkeit, bequeme Zugänglichkeit für Instandhaltung, Möglichkeit der Abtrennung eines Teilsystems für die Instandhaltung ohne Beeinträchtigung der Funktion des Gesamtsystems, Prüf- und Eichmöglichkeit, Ersatzteilhaltung u. a.), der Automatisierung (Was sollte automatisiert werden? Durchgängige Automatisierung ist nicht anzustreben. Bei seltenen Vorgängen, wie z. B. dem Turbinenanfahren wird es nicht für sinnvoll erachtet.) und des Rechnereinsatzes. Interessant ist auch die Andeutung von Entwicklungstendenzen in den Grundkonzeptionen, z. B. der Entwicklung des Sicherheitskonzeptes von einer reinen ,,schwarz/weiß"-Klassifikation zu einer ,,grau"-Klassifikation, die Alarme entsprechend ihrer Bedeutung untersetzt, um unnötige Aktionen des Sicherheitssystems zu vermeiden.

Obwohl das Handbuch in erster Linie für Projektanten geschrieben wurde, kann es auch all denen empfohlen werden, die mit dem KKW-Betrieb, mit der Ausbildung von KKW-Fachleuten, mit Sicherheitsproblemen und mit der Entwicklung von Meß-, Diagnose- und Automatisierungsmitteln für KKW zu tun haben. Viele der Betrachtungen bezüglich Sicherheit, Instandhaltung usw. sind auch auf konventionelle industrielle Großanlagen übertragbar, so daß auch Fachleute aus dieser Richtung das Buch mit Gewinn lesen werden.

P. Liewers

Thin-Film and Depth-Profile Analysis

Hrsg.: H. Oechsner. Berlin – Heidelberg – New York – Tokyo: Springer 1984. 225 S., 99 Abb., zahlr. Tab., 515 Lit. (Topics in Current Physics Vol. 37); geb. DM 72,–

Die technische Bedeutung dünner Schichten nimmt schnell zu. Das erfordert die verläßliche Kenntnis des Zusammenhangs von Bildungs- und Strukturparametern sowie von Eigenschaften solcher Schichten. Speziell die Lateral- und die Tiefenverteilung chemischer Elemente spielen eine

Rolle. Zur Kenntnis der Methoden zur Tiefenabtastung leistet die Monographie einen zusammenfassenden wenn auch nicht unbedingt systematischen Beitrag. Im Vordergrund steht die Verbesserung der Quantifizierbarkeit und der Nachweisgranze für die verschiedenen Elemente in einem schmalen Tiefenintervall und die Reduzierung der Ausdehnung eines solchen Intervalls bei möglichst genauer Festlegung seiner absoluten Tiefe.

In der von *H. W. Werner* (Philips-Eindhoven) gegebenen Übersicht zu den Strahl- und Beugungstechniken wird die ganze Reihe von Acronymen vorgeführt, die manche Autoren für zweckmäßig erachten, indem die Wahl von ein/zwei speziellen Parameterwerten bei eingeführten Techniken ihnen anscheinend hinreichende Gründe dafür liefert. Nichts gegen alteingeführte Bezeichnungen wie AES oder ESCA, aber muß es denn sein, daß sich z. B. hinter ISS, HEIS, LEIS und RBS eigentlich nicht anderes als die alte klassische *Rutherford*-Streuung versteckt? Die *Werner*schen Übersichten erscheinen in ihren Daten zum Teil etwas willkürlich und überholt (Literatur von 1980). Beispiel: Warum sollten sich die Strahldurchmesser bei PIXE und *Rutherford*-Streuung so stark unterscheiden, wie in Tab. 2.4. (S. 9) angegeben? Hierfür gibt es physikalisch und meßpraktisch keinen Grund. Daß die für eine Tiefenprofilabtastung so nützlichen Kernresonanzreaktionen fast vollständig unerwähnt bleiben, ist bedauerlich.

Eine interessante Variante mit ausgezeichneter Quantifizierbarkeit bei Anwendung sehr niedriger Ionenenergien ist die Sekundär-Neutralteilchen-Massenspektrometrie, die *H. Oechsner* (Univ. Kaiserslautern) vorgestellt. Sie kommt dem Ideal der Oberflächenanalytik — geringe Informationstiefe — anscheinend am nächsten.

Die Sputtertechnik als Hilfsverfahren zur Schichtabtragung nimmt in dem Buche den größten Raum ein. Der ausführliche aber nicht leicht in die Praxis umsetzbare Beitrag von *U. Littmark* und *W. O. Hofer* (KFA Jülich) beginnt mit einem mehr phänomenologischen Überblick zu dem beim Sputtern wesentlichen Rückstoß-„mixing" und geht sodann zur Diskussion dazu vorhandener theoretischer Modelle über. Die Bedeutung dieses ausgezeichneten Artikels ergibt sich aus seiner aktuellen (Literatur bis 1983) und präzisen Übersichtsdarstellung des „mixing" als der eigentlichen inhärenten Begrenzung des Sputterns als Methode des mehr oder weniger geordneten Schichtabtragens. Im Unterschied zu früheren Darstellungen dieses Gebiets legen die Autoren Wert darauf, die vollständigen Delokalisationsverteilungen und nicht nur einzelne Verteilungsparameter zu diskutieren. Einige weitere Beiträge behandeln speziellere Aspekte besonders der *Auger*-Elektronenspektroskopie und der Sekundärionen-Massenspektroskopie.

Jeder Physiker oder Chemiker in Forschung und Industrie, der sich aktiv oder passiv mit Oberflächenanalytik in Kombination mit Schichtabtragung befaßt, vermag aus dem Buche Nutzen zu ziehen, indem es ihm dicht gedrängt Möglichkeiten und Grenzen einer ganzen Reihe von Methoden und Varianten vorführt und im Hinblick auf bestimmte Anwendungsfälle wertvolle Vergleiche ermöglicht.

G. Brunner

The Handbook of Environmental Chemistry Vol. 3, Part C: Anthropogenic Compounds

O. Hutzinger (ed.). Berlin — Heidelberg — New York — Tokyo: Springer-Verlag 1984. 220 S., 31 Abb., 62 Tab.; geb. DM 138,—; US$ 51.50

Von dem gut eingeführten Handbuch der Umweltchemie liegt nunmehr der Teil C des 3. Bandes vor, der sich — wie auch bereits die Teile A und B — mit Verbindungen beschäftigt, die der Mensch in die Umwelt einträgt, die anthropogenen Verbindungen. So werden einerseits organische Verbindungen behandelt, die zwar für bestimmte Zwecke von der chemischen Industrie hergestellt werden, die unsere Umwelt aber als Gifte aufnimmt und abbauen muß. Andererseits können chemische Verbindungen an die Umwelt abgegeben werden in Konzentrationen, die ebenfalls Giftwirkungen zur Folge haben, wie es das Beispiel Thallium zeigt. In beiden Fällen muß die Biosphäre mit diesen Verbindungen „fertig" werden und sie so weit ab- oder umbauen, bis eine gewisse Ungiftigkeit erreicht wurde, die aber leider auch von Schadenserscheinungen begleitet ist.

In dem vorliegenden Teil werden aromatische Amine (40 S.), Phosphatester (25 S.), Phthalsäureester (75 S.) und Thallium (71 S.) behandelt. Es ist sehr nutzerfreundlich, auch wenn jeder Abschnitt eine ausführliche Erläuterung der verschiedenen analytischen Methoden beinhaltet, womit der Handbuchcharakter besonders deutlich wird. Daneben sind allgemeine Angaben über das chemische und physikalisch-chemische Verhalten der Stoffe zu finden. Hauptanliegen ist naturgemäß das Verhalten der Stoffe in unserer Umwelt, insbesondere ihre Giftwirkung und ihr bioorganischer Abbau (Biodegradation). Vom Tl und seinen Verbindungen werden die Methoden der RFA und der NAA im Vergleich mit den anderen spurenanalytischen Verfahren vorgestellt.

Auch dieser Teilband zeichnet sich durch eine Fülle von Literaturhinweisen aus (insgesamt 1126 Zitate), die eine hohe Aktualität haben (allein auf die Jahre 1980—1983 entfallen 362 Zitate ≙ 32,2%). Dies zeigt, daß wir bereits viel Wissen auf diesem Gebiet angehäuft haben; trotzdem sollte der Einsatz bestimmter Stoffe, die nur einem erhöhten Gebrauchswert dienen, nicht als alleiniges Kriterium dienen. Unsere Umwelt muß diese exotischen Stoffe auch unbeschadet aufnehmen können, wenn sich nicht bleibende Folgen einstellen sollen. Dem Umweltchemiker werden auch mit diesem sehr daten- und faktenreichen Buch wertvolle Hinweise und Beweise gegeben, wie sehr es not tut, mit unserer Umwelt sorgsam umzugehen.

H. Hübner

Spin Polarization and Magnetic Effects in Radial Reactions

K. M. Salikhov, Yu. N. Molin, R. Z. Sagdeev, A. L. Buchachenko. Ed.: Yu. N. Molin. Budapest: Akademiai Kiado 1984. 419 S., 153 Abb., zahlr. Tab., Sachwortverzeichnis, Lit. (aus dem Russischen); geb. M 90,—

Magnetische Effekte und Spinpolarisation, wie sie bei Radikalreaktionen in kondensierter Phase auftreten, basieren auf einem gemeinsamen physikalischen Mechanismus. Voraussetzung für das Auftreten dieser Erscheinungen ist die Existenz von Radikalpaaren im „Käfig" umgebender Moleküle. Die Lebensdauer von Radikalpaaren im „Käfig" schwankt zwischen 10^{-11} und 10^{-8} s. Sie ist entweder durch die Reaktion der Partner des Radikalpaares miteinander (Rekombination, Disproportionierung) oder durch das Entkommen eines Radikals aus dem „Käfig" begrenzt. Es ist seit langem bekannt, daß die Lebensdauer dieser Radikale durch ihre Reaktivität, ihre Diffusionskoeffizienten, elektrostatische Wechselwirkungen usw. bestimmt wird. Obwohl die Erzeugung und die Rekombination von Radikalpaaren unter Spinerhaltung erfolgen, können Spin-Übergänge in Radikalpaaren durch magnetische (*Zeeman-*, Hyperfein-) Wechselwirkungen ausgelöst werden. Bemerkenswert ist dabei die Tatsache, daß diese Effekte durch Energieänderungen hervorgerufen werden, die um Größenordnungen unter der thermischen Energie liegen.

Bereits 1967 wurde an Reaktionsprodukten von Radikalen Kernspinpolarisation nachgewiesen (Chemically Induced Dynamic Nuclear Polarization — CIDNP). Ein Jahr später fand man Spinpolarisation in ESR-Spektren von Radikalen (Chemically Induced Dynamic Electron Spin Polarization — CIDEP). Nach 1976 wurde zur experimentellen Untersuchung des Einflusses äußerer und innerer magnetischer Felder auf die Geschwindigkeitskonstanten von Radikalreaktionen neben der HF-Spektroskopie auch die optische Kurzzeitspektroskopie eingesetzt. Seitdem hat sich die Untersuchung magnetischer Effekte an Radikalreaktionen zu einem schnell expandierenden Gebiet der modernen physikalischen Chemie entwickelt. Die Autoren der vorliegenden Monographie, die selbst maßgeblich zur Entwicklung auf diesem Gebiet beigetragen haben, bieten den umfangreichen Stoff in logisch gegliederter Form. Auf eine phänomenologische Beschreibung der Reaktion von Radikalpaaren und ihrer Singulett-Triplett-Übergänge (*Yu. M. Molin*) folgt ein Kapitel über die Theorie der Radikalrekombination (*K. M. Salikhov*), das zunächst in kurzer Form die spinunabhängige Radikalrekombination behandelt, um danach ausführlich auf die Singulett-Triplett-Übergänge in Radikalpaaren einzugehen. Dieser Teil des Kapitels wird ergänzt durch einen Abschnitt über mathematische Hilfsmittel der Theorie. Besonders hervorzuheben ist die Vielzahl von Berechnungsbeispielen, die die einzelnen möglichen Fälle gut illustrieren. Im anschließenden Kapitel (*K. M. Salikhov*) wird die Theorie magnetischer Effekte in Radikalreaktionen ausführlich beschrieben. Dabei werden alle wesentlichen Modelle (Stand etwa 1980) zur Beschreibung der Radikalrekombination in starken und schwachen äußeren magnetischen Feldern und der magnetische Isotopieeffekt berücksichtigt.

In den folgenden Kapiteln werden dann experimentelle Befunde behandelt, die jeweils auf die Beeinflussung von Radikalrekombinationsreaktionen durch Spin-Auswahlregeln zurückgeführt werden. *Yu. N. Molin* und *R. Z. Sagdeev* diskutieren den Einfluß magnetischer Effekte auf die Rekombination neutraler Radikale, Radikalionen und Tripletts. Besonders hervorzuheben ist die ausführliche Beschreibung von Technik und Aussagemöglichkeiten des optischen Nachweises von Radikalpaaren durch ESR (ODESR).

Anschließend behandeln *R. Z. Sagdeev* und *A. L. Buchachenko* den magnetischen Isotopieeffekt und ausgewählte CIDNP-Experimente. Breiten Raum nimmt dabei die Diskussion der Aussagemöglichkeiten von CIDNP-Experimenten zur Aufklärung komplizierter Reaktionsmechanismen ein.

Der Wert dieser Monographie liegt in der geschlossenen Darstellung der Theorie und der wesentlichen experimentellen Tatsachen des Einflusses von Spin-Auswahlregeln auf die Kinetik chemischer Reaktionen. Der Stand der Literatur wird etwa bis 1980 berücksichtigt. Wie die Autoren richtig voraussagen, hat sich dieses Gebiet seitdem stürmisch entwickelt. Hervorzuheben sind insbesondere experimentelle Entwicklungen wie die zeitaufgelöste optisch detektierte ESR und die breiten Möglichkeiten der Radikalerzeugung durch leistungsstarke UV-(Excimer)Laser. Hier ist der Zeitverzug zwischen Abschluß des Manuskripts (etwa 1980) und dem Erscheinungsjahr der Monographie deutlich spürbar.

Das Buch kann Physikern, Chemikern und Biologen empfohlen werden, die sich mit CIDNP und CIDEP befassen. Als Interessenten kommen außerdem Naturwissenschaftler in Frage, die sich mit der Aufklärung von chemischen Reaktionsmechanismen und der Kinetik von Elementarreaktionen beschäftigen.

R. Mehnert

Physikalische Chemie, Gesamtausgabe

G. M. Barrow, bearb. von G. W. Herzog, 6. berichtigte Aufl.; Wien — Braunschweig/Wiesbaden: Bohmann Verlag u. Friedr. Vieweg & Sohn 1984. 941 S., 447 Abb., 124 Tab.; geb. DM 88,—

Die physikalische Chemie spielt heute eine zentrale Rolle in der Forschung und damit auch in der Ausbildung. Das zunehmend große Gebiet hat sich seit den Anfängen um die Jahrhundertwende in spezielle Einzeldisziplinen aufgespalten: physikalisch-organische Chemie, physikalisch-anorganische Chemie, physikalische Biochemie, physikalische Geochemie u. ä. m. Das vorliegende Lehrbuch ist sozusagen die klassische physikalische Chemie, also die Thermodynamik, Atomistik, Kinetik und Elektrochemie. Zum Unterschied zur vorhergehenden Auflage hat sich der Verlag entschlossen, die bisherigen 3 Teilbände wieder in einer Gesamtausgabe erscheinen zu lassen — dieser Entschluß ist zu begrüßen, hängen doch die einzelnen Abschnitte sehr eng zusammen.

Die Rezension einer 6. Auflage kann im wesentlichen aufs neue feststellen, daß sich der Aufbau, die Art der Darstellung und die Auswahl aus dem großen Bereich der physikalischen Chemie sehr bewährt haben und daß das Buch auch künftig bei einem dankbaren Leserkreis eine gute Aufnahme finden wird. Einige Anmerkungen scheinen aber geboten: So wurde eine ausführlichere Darstellung der Isotopenaustauschreaktion $D + H_2 \rightarrow DH + H$ im Zusammenhang mit der Theorie des Übergangszustandes neu aufgenommen. Diese Theorie — gerade jetzt 50 Jahre alt geworden — hat für den Chemiker vielfach ihre fruchtbare Anwendbarkeit bewiesen. Nur diese gewählte Reaktion, so einfach sie auch erscheinen mag, bedarf doch der Tunnelkorrektur und der Berücksichtigung von Anharmonizitätsgliedern, so daß ein Vergleich zwischen berechneten und gefundenen Werten nicht sehr überzeugend wirken muß. Hier wären vorgeführte Rechnungen an anderen Beispielen (Tabelle 22.6 zur Auswahl) didaktisch günstiger. Daß man für Nuklide auch die Angabe der Massenzahl rechts oben zuläßt, dürfte bei der Schreibung einer einzigen chemischen Reaktionsgleichung mit Ionenladungen zu Schwierigkeiten führen (S. 296, Teil I). In der Isotopentabelle (Anhang) wird D in Klammern gesetzt, was auf Instabilität dieses Nuklids hindeuten soll. Außerdem sind in dieser Tabelle so wichtige Nuklide wie T, ^{14}C, ^{36}Cl u. a. nicht mit aufgenommen, wobei diese kosmogenen Nuklide als in der Natur vorkommend doch genannt werden sollten. Ferner sollte einmal der konsequente Durchbruch zu den Bezeichnungen Ethan, Ethen, Benzen, Toluen u. a. gewagt werden, da der Student wohl durch Vorlesungen und Lehrbücher für organische Chemie mit diesen richtigen Bezeichnungen bereits vertraut sein müßte (IUPAC-Empfehlung).

Auch diese Neubearbeitung des „Barrow“ wird eine gute Aufnahme finden: dabei wird die gute Anschaulichkeit und leichte Verständlichkeit des dargebotenen Stoffes dem Leser und Nutzer helfen.

H. Hübner

Mikroelektronik für Führungskräfte
Eine Anleitung für Einsteiger

Hrsg.: M. Becker und *H. Tarschisch. Zürich: Verlag Industrielle Organisation 1984. 416 S., zahlr. Abb., Tab. u. Lit.; geb. SFr* 88.—

Im Rahmen eines Impulsprogramms, das der Schweizerische Bundesrat unter der Bezeichnung „Maßnahmen zur Milderung der wirtschaftlichen Schwierigkeiten“ vor einigen Jahren beschlossen hat, findet sich auch ein Komplex zur Förderung der Mikroelektronik-Technologie. Er ist auf die Erhaltung der internationalen Konkurrenzfähigkeit durch Anwendung der Mikroelektronik in Produkten der Schweizer Klein- und Mittelbetriebe gerichtet. In einer beispielhaften Initiative des Bundesrates erhielt das Betriebswissenschaftliche Institut der Eidgenössischen Technischen Hochschule in Zürich den Auftrag zu einer umfassenden mikroelektronischen Software-Beratung. Dabei ist ein Seminar „Mikroelektronik für Führungskräfte“ wesentlicher Bestandteil. Sein Stoff liefert den Inhalt des vorliegenden Buches.

Zehn versierte Fachleute, unter ihnen erfahrene Industrie-Techniker, sind die Autoren folgender Kapitel: Einführung, Grundlagen, Praxis des Einsatzes der Mikroelektronik, Erfahrungsberichte und praktische Hinweise, Auswirkungen und gesellschaftliche Aspekte sowie eines „Kleinen Lexikons der Mikroelektronik“.

Das Textverständnis erfordert der Zielsetzung nach kaum besondere technische Vorkenntnisse. Der Text wurde ausgesprochen weit aufgegliedert. Sehr viele eingestreute Abbildungen, Skizzen und Tabellen kommen dem Bemühen des Lesers um den an sich nicht ganz einfachen Stoff sehr entgegen. Das Ergebnis: Auch wenn man nicht zur Zielgruppe des Industriemanagers gehört, liest man das Buch voller Interesse von Anfang bis Ende durch. Besonders die beiden Anfangskapitel, und da wiederum die Darstellung des Software-Engineering, verdienen nach Ansicht des Rezensenten ein hohes Lob. Hier wurde sehr bewußt im Sinne der spezifisch-schweizerischen Gesichtspunkte der Klein- und Mittelindustrie und der relativen Bevorzugung der Dienstleistungen dem allgemein beobachtbaren Trend der Verstärkung des Software-Anteils bei der Mini- und Mikrocomputeranwendung Rechnung getragen. Der Leser wird in die Lage versetzt, nicht nur zu verstehen, sondern fachlich mitzudiskutieren und vor allem zu entscheiden. Ihm wird z. B. klargemacht, wieso heute die klare Strukturiertheit im Sinne von *Nassi-Shneiderman* oder *M. Jackson* der komplexen und vermaschten „GOTO“-Programmierung vorzuziehen ist, und wie sich das letztlich im Aufwand/Nutzen-Verhältnis niederschlägt, wie das Feld aus Produkten verschiedener Marktattraktivität und unterschiedlicher Wettbewerbsstärke zweckmäßig abzudecken ist oder warum und wie ein Pflichtenheft zur Produktentwicklung zu erarbeiten ist. Bei den Ausführungen über konkrete Programmierung wird übrigens der Mikroprozessor 8080 (INTEL) als Beispiel verwendet.

Das ein wenig gesellschaftsphilosophische Schlußkapitel über die Dritte Industrielle Revolution ist gleichfalls sehr interessant zu lesen, aber in seinem Aussagewert natürlich etwas problematisch. Durch die Etablierung von Einbis Zweimann-Kleinstunternehmen des Software-Engineering wird man die Arbeitslosigkeit in den kapitalistischen Ländern wohl kaum beeinflussen, aber das sieht wohl der Kapitelautor im Ernst auch selbst sicherlich kaum anders.

Es bleibt zu wünschen, daß solche anerkennenswerten Initiativen, wie sie zu diesem Buch geführt haben (s. o.!), auch in unserem Lande noch mehr Platz greifen, denn wenn man an die Rohstoffsituation oder das Bemühen um marktbedingtes intelligenz- bzw. dienstleistungsbetontes Herangehen der Exportindustrie erinnern darf, so gibt es durchaus bestimmte Parallelen zur schweizerischen Situation in den industriellen Randbedingungen. Zumindest mögen viele für die Profilgestaltung unserer mikroelektronischen Industrie Verantwortliche Gelegenheit nehmen, sich mit dem Inhalt dieses bemerkenswerten Buches vertraut zu machen.

G. Brunner

Atmosphäre und Umwelt
Chemische Prozesse — Menschliche Eingriffe

P. Fabian, Berlin — Heidelberg — New York — Tokyo: Springer-Verlag 1984. X, 115 S., 30 Abb., 5 Tab.; geheftet DM 24,—; approx. US$ 9.40

Fragen der Luftverschmutzung, ihre Ursachen und ihre Folgen, finden heute ein hohes Interesse bei vielen Menschen. Unsere Atmosphäre, wie sie uns jetzt umgibt, zeigte in der geologischen Entwicklung nicht immer die uns bekannte Zusammensetzung. Vor Milliarden Jahren, im Präkambrium, waren von dem so lebensnotwendigen Sauerstoff nur Spuren vorhanden, und durch die Wirkung der Photosynthese erreichte er langsam den heutigen Anteil von rund 21 Vol-%. Damit war auch die zur Besiedelung des Festlandes notwendige Bildung des Ozonschildes möglich, der die lebensfeindliche UV-Strahlung der Sonne ausfiltert. Eine Reihe von Spurengasen im ppm- und ppb-Bereich (auch ppm-Gase genannt) zirkulieren in unserer Atmosphäre und können durch die Tätigkeit des Menschen erhöhte Konzentrationen erreichen, die im natürlichen Gleichgewicht nicht ohne weiteres abgebaut werden können bzw. die nicht ohne Auswirkungen auf dieses Gleichgewicht sind.

Der Autor folgt dieser angedeuteten Reihenfolge auch in der kleinen Schrift: Entstehung des Sauerstoffs in der Atmosphäre und Bildung des Ozonschildes, photochemische Reaktionen in der Troposphäre und Auswirkungen bestimmter ppm-Gase auf das natürliche Gleichgewicht, insbesondere auf Veränderungen der Ozonkonzentration in der Stratosphäre. Einige Ungenauigkeiten haben sich bei dieser knappen Darstellung eingeschlichen. Die Prokaryonten als primitive Einzeller haben zwar keine Zellkernstruktur, keine echten Zellkerne, aber doch wohl eine einfache, gering kompartmentierte Zellstruktur (S. 14). Oxydation und Oxidation werden uneinheitlich verwendet (S. 14, 16, 82). Das $^{13}C/^{12}C$-Verhältnis oder besser der $\delta^{13}C$-Wert in organischen Substanzen ist nicht um 25 ‰ vom Umwelt-CO_2 verschieden, sondern nur rund 18 ‰; die 25 ‰ beziehen sich auf den Carbonat-Kohlenstoff (S. 21). Wäre es nicht didaktisch zweckmäßiger, die *Dobson*-Einheit für stratosphärisches Ozon in ppb-Angaben umzurechnen (S. 55)? Es ist doch für den Leser nicht leicht einzusehen, daß eine Konzentrationsangabe oder ein Mischungsverhältnis in atm-cm erfolgen kann. Man sollte auch die Massenzahlangabe eines Elements nicht in der Form ^{90}Strontium verwenden (S. 78); entweder ^{90}Sr oder Strontium-90.

Insgesamt dürfte die Schrift die angestrebte Zielgruppe von Lesern erreichen — den interessierten Laien, der manches von der vielerorten diskutierten Verschmutzung der Atmosphäre hört und gern Näheres darüber wissen möchte. All diesen Interessenten kann die Schrift empfohlen werden, um eine Wissenslücke zu füllen und um sein eigenes Tun selbst einzuordnen.

H. Hübner

Licht vom Rande der Welt
Das Universum und sein Anfang

R. Kippenhahn. Stuttgart: Deutsche Verlags-Anstalt 1984. 352 S., 88 Abb.; geb. DM 39,80

Giordano Bruno, italienischer Renaissancephilosoph und von den Schriften des *Kopernikus* begeistert, hat einmal gesagt, die Welt sei eine ewige Unendlichkeit. Wenn wir auch heute die Größe des Raumes und die Länge der Zeit in Exponentialform leicht darstellen und damit rechnen können, so haftet diesen Dimensionen doch die Unvorstellbarkeit an.

Einen anschaulichen und modernen Zugang in die Welt dieses ewig Unendlichen bietet hier der Astrophysiker Professor *Rudolf Kippenhahn*, und der aufmerksame Leser folgt ihm gern. Mit leichter Hand führt er uns — in 13 Kapiteln — von der Anatomie der Milchstraße über Urknall und Schwerkraft in das Reich der Nebel, der Radiogalaxien, zu den Quasaren und Schwarzen Löchern. Dabei bedient er sich um der Anschaulichkeit willen auch einer humorvollen Darstellung, wie beispielsweise in dem mehrfach in Anspruch genommenen Bild einer Welt von Flachmännern, die nur zweidimensional zu sehen und damit zu denken vermögen.

Eine moderne Kosmologie und doch schon wieder veraltet? Gerade im Erscheinungsjahr dieses Buches fand in Leipzig die XXII. Internationale Konferenz über Hochenergiephysik statt, auf der u. a. für das vorliegende Buch zumindest zwei neue Erkenntnisse vorgetragen und diskutiert wurden: die kosmische Inflation und die gegenwärtig am wahrscheinlichsten anzunehmende Neutrinomasse $20 < M_\nu < 45$ eV oder $(4 \ldots 9) \cdot 10^{-5}$ Elektronenmassen M_e (*V. A. Lubimov*, Moskau). Gerade das Modell der Inflation kann ein noch offenes Problem der Kosmologie lösen, nämlich die gegenwärtige Verteilung der Galaxien und Galaxiencluster im sichtbaren Teil unseres Universums. Auf S. 305 wird gesagt, daß sich das Deuterium vom Wasserstoff — oder wohl besser vom Protium chemisch nicht unterscheiden würde. Wenn auch die Unterschiede im chemischen und physikalisch-chemischen Verhalten dieser beiden Nuklide des Wasserstoffs nur geringfügig sind, so sind sie doch beobachtbar und auch erklärbar (Isotopieeffekte), und sie sind eine der Ursachen für die Schwierigkeiten, aus Spektren der interstellaren und planetaren Verbindungen auf den D-Gehalt zu schließen.

Alle Leser, die sich von dem erfahrenen Astrophysiker in die weite Welt des Sternenhimmels führen und verführen lassen, werden ihm für diese wohlgelungene Darstellung dankbar sein. Die Auswahl der Bilder und die Wahl der Vereinfachungen helfen dem Interessierten sehr, und er wird nach Beenden des Buches ein klein wenig erahnen können, was *Giordano Bruno* mit der ewigen Unendlichkeit gemeint haben könnte.

H. Hübner

Metallformung
Geschichte — Kunst — Technik

K.-D. Lietzmann, J. Schlegel, A. Hensel. Leipzig: VEB Deutscher Verlag für Grundstoffindustrie 1984. 256 S., 240 teils farb. Abb., 98 Bildmarginalien; Leinen M 65,—

In den letzen Jahren sind in verstärktem Maße zusammenfassende Darstellungen von Technik- bzw. Wissensgebieten erschienen, die sich durch Verständlichkeit, Übersichtlichkeit, hohen Informationsgehalt und nicht zuletzt eine anspruchsvolle Gestaltung auszeichnen. Sie sind keine Lehrbücher, verzichten meist gänzlich auf mathematische, chemische oder physikalische Formulierungen und vermitteln trotzdem und deshalb sowohl dem Fachmann als auch dem wißbegierigen Laien weitere und neue interessante Kenntnisse über das jeweilige Gebiet.

In diese Gruppe von Büchern fällt das Werk „Metallformung", dessen Untertitel „Geschichte — Kunst — Technik" bereits ahnen läßt, was den Leser erwartet.

In den fünf Hauptabschnitten:

Zur Frühgeschichte der Metallformung, Die Metallformung im Mittelalter, Ausgewählte Techniken der Metallformung, Die Metallformung vom Mittelalter bis zum Beginn des 20. Jahrhunderts und Die Metallformung in Gegenwart und Zukunft,

wird ein Abriß der Metallformung von ihren ersten Anfängen bis zur Gegenwart gegeben. Dabei finden die historischen Fakten, die technischen Belange und auch künstlerische Aspekte ihren angemessenen Platz. So wird neben der technischen Entwicklung und dem Betrieb moderner Großumformanlagen mit einer Jahresproduktion von vielen Tausend Tonnen auch auf die Kunstfertigkeit der Handwerker und ihre sowohl zweckentsprechenden als auch formvollendeten Erzeugnisse eingegangen. Die oft jahrhundertealte Tradition und die Entwicklung der mannigfachen Verformungstechniken für Metalle wie die Gold- und Silberschmiedekunst, die Blattgoldherstellung, die Toreutik, die Eisenschmiedekunst, die Waffenschmiedekunst, das Prägen von Metallen sowie die Verfahren des Schmiedens, Walzens und Ziehens zur Herstellung von Formstahl, Blechen, Bändern und Drähten werden kurzweilig und mit hoher Sachkenntnis dargestellt. Ein umfangreiches, äußerst instruktives und teils farbiges Bildmaterial veranschaulicht dem Leser auch weniger bekannte alte und neue Verformungstechniken und die handwerklichen und künstlerischen Leistungen der Metallformer. Zeittabellen zur Entwicklung der Metallformung und ein Quellenverzeichnis der Zitate und der Literatur schließen das sehr gelungene Werk ab.

Unser heutiges Leben, ob in der privaten Sphäre oder im Beruf, ist ohne Metalle und damit ohne die Metallformung nicht denkbar. Es erscheint daher auch nicht abwegig, einem breiten Kreis von Menschen praktisch aller Berufe, ob jung oder alt, ein solches, für viele zunächst scheinbar abseits gelegenes Fachgebiet näherzubringen. Das vorliegende Buch ist dazu ohne Zweifel geeignet.

J. Holzhey

Jahrbuch der Atomwirtschaft 1985

Hrsg.: W. D. Müller, R. Hossner. Düsseldorf/Frankfurt: Verlagsgruppe Handelsblatt 1985. 468 S., 112 Abb., 124 Tab., 135 Lit.; geb. DM 56,—

Das Jahrbuch 1985 wurde entsprechend der Grundkonzeption der früheren Auflagen auf dem Stand August 1984 überarbeitet und ergänzt. Im Abschnitt Perspektiven werden die Schwerpunkte der Entwicklung der Kernenergie 1985 im Vergleich zur Energieerzeugung auf der Basis Kohle dargestellt. Der Anteil der Kernenergie an der Gesamtenergieerzeugung wird Anfang der 90er Jahre in der BRD 35%, in Frankreich 75% betragen. Es wird auf die Probleme der Umweltbelastung durch SO_2 und NO_x bei der Stromerzeugung aus Kohle und damit auf die steigende Wirtschaftlichkeit der Kernkraftwerke ebenso wie auf die Bedeutung der Wiederaufbereitung hingewiesen. Zukünftige F/E-Arbeiten konzentrieren sich auf Probleme der Reaktorsicherheit, der Urananreicherung mit Laser, der Wiederaufbereitung und Rezyklisierung von Kernbrennstoffen und der sicheren Endlagerung radioaktiver Abfälle (*M. Popp*). Das Atomenergierecht speziell in der BRD wird von *E. Ziegler* und *W. Boucsein* behandelt. Die Kernbrennstoffversorgung wird von *K. Braatz* und *H. J. Dibbert* als günstig und gesichert angesehen und mit umfangreichen Übersichten der Uranreserven in der Welt, der Darstellung der Produktionskapazitäten und der Preisentwicklung belegt. Zur Endlagerung radioaktiver Abfälle in der BRD werden von *H. Röthemeyer* die Projekte Gorleben und die Schachtanlage Konrad vorgestellt. *L. Wiesner* behandelt Probleme der Strahlenwirkungen und Strahlenanwendungen im biologischen Bereich, dabei besonders die Erzeugung von DNA-Defekten und ihre Reperatur, das umstrittene Problem der Hormesis durch ionisierende Strahlung und die Strahlenstimulation von Nutzpflanzen, die besonders in den Ländern des RGW zur Anwendung kommt.

Der Adressenteil informiert über Organisationen und Forschungseinrichtungen in der BRD, die sich mit Problemen der Nutzung der Kernenergie befassen.

Der Abschnitt Informationen enthält Angaben zum Weltmaßstab, zu Europa und 52 Ländern nach sachlichen Gesichtspunkten wie z. B. Kernkraftwerke, Reaktoren, Atomprogramme, Meß- und Regeltechnik, Radioisotope, Sicherheit, Strahlenschutz, Forschung, Ausbildung oder internationale Abkommen.

Die umfangreiche Zusammenstellung von Daten ergibt eine internationale Übersicht (Stand August 1984) von Kernkraftwerken im Betrieb, im Bau bzw. in der Projektierung, Angaben über Reaktorhersteller, über die Kernenergie in der Energiewirtschaft, über Betriebsergebnisse von Kernkraftwerken wie Bruttostromerzeugung und Lastfaktor 1983, die Verfügbarkeit und Angaben zur Wiederaufbereitung bestrahlter Kernbrennstoffe.

Das vorliegende Jahrbuch ist ein wertvolles Ergänzungsnachschlagewerk zu den vorhergehenden Auflagen, da es auf Grund der Aktualität den Fortschritt der Kernenergetik und Atomwirtschaft erkennen läßt. Es gehört daher zum Standard jeder Bibliothek von Einrichtungen, die sich mit wissenschaftlichen, technologischen oder ökonomischen Problemen der Energiewirtschaft bzw. der Isotopentechnik befassen.

H.-G. Könnecke

Nuclear Heat Application
Proc. of a Technical Committee Meeting and Workshop on Nuclear Heat Application organized by the IAEA and held in Cracow, 5–9 December 1983

Vienna: IAEA 1984. 419 S., div. Abb. u. Tab., 117 Lit. (Proc. Series STI/PUB/679); kart. ÖS 820,–

Die IAEA-Tagung 1983 in Krakau ist Ausdruck des Bemühens, die international bereits seit längerer Zeit durchgeführten Arbeiten zur nuklearen Wärmeversorgung im Niedertemperaturbereich bis 250 °C sowie im Hochtemperaturbereich von 500 °C–1000 °C zusammenzufassen. 64 teilnehmende Wissenschaftler aus 16 Ländern stellten mit 32 Beiträgen in 5 Sessionen die differenzierten Voraussetzungen und Erfahrungen ihrer Länder vor. Die DDR war durch drei Teilnehmer vertreten. Nach einleitenden Ausführungen zur Analyse des Verbraucherpotentials für Heiz- und Prozeßwärme in der BRD und der ČSSR sowie Abschätzungen für die Städte Zagreb und Budapest wurden in den folgenden zwei Abschnitten technische Lösungen und Anwendungsbeispiele für nukleare Wärmequellen vorgestellt. Dazu zählen vornehmlich die Projekte AST-500, AST-300 (UdSSR), SECURE (Schweden), SLOWPOKE (Kanada) und das KWU-Projekt (BRD), die ausschließlich der Heizwärmeversorgung dienen. Eindeutig erkennbar ist dabei die Tendenz zum Einsatz heterogener, naturumlaufgekühlter Leichtwasserreaktoren für die Niedertemperatur-Wärmebereitstellung. Dies wird ebenfalls durch die Schlußfolgerungen von *Seifritz* (Würenlingen) nach Untersuchungen an einem homogenen Reaktorsystem mit einer Leistung von 10 MW_{th} bestätigt. Den führenden Stand sowjetischer Arbeiten auf dem Gebiet der nuklearen Heizwärmeversorgung bringen die Beiträge von *I. N. Sokolov* (Kurtschatov-Institut, Moskau) anschaulich zum Ausdruck. Im weiteren Verlauf nahmen Beispiele für die Wärmeauskopplung aus KKW verschiedener Bauart sowie die Projekte von Hochtemperaturreaktoren der USA, der BRD, Japans und der UdSSR breiten Raum ein. Interessante Einsatzaspekte gasgekühlter Hochtemperaturreaktoren in Bereichen der chemischen Industrie, der Metallurgie, der Kohlevergasung und der Wasserstoffproduktion, wurden im Zusammenhang mit einer ökonomischen Wertung des Anlagenbetriebes bereits bestehender Pilotanlagen von Vertretern dieser Länder vorgetragen. Hohe Investitionskosten und eine komplizierte Technologie verhindern gegenwärtig den Einsatz derartiger Anlagen, wobei jedoch auch denkbare Weiterentwicklungen in den kommenden Jahren kaum deutliche Vorteile gegenüber kostengünstiger Elektroenergie aus schnellen Brutreaktoren als Alternative erwarten lassen. In einem ergänzenden Abschnitt Ökonomie gaben drei Beiträge Informationen zu ökonomischen Randbedingungen bei der Bewertung der nuklearen Wärmeversorgung im Niedertemperaturbereich. Die abschließend angefügten Strahlenschutzregelungen und nationalen Standards (UdSSR) für den Einsatz nuklearer Wärmequellen unterstreichen das außerordentlich hohe Sicherheitspotential dieser Anlagen. Hinweise für zulässige Dosisleistungen und Kontrollmechanismen in den Verbraucherkreisläufen ergänzen die Aussagen.

Das Buch kann wissenschaftlich und ingenieurtechnisch Tätigen in den Bereichen Kernenergetik und Energieversorgung sowie Lehrkräften und Studenten energetischer Studienrichtungen zur Information und Weiterbildung empfohlen werden.

J. Sydow

EXAFS and Near Edge Structure III
Proc. of an International Conference, Stanford, CA, July 16–20, 1984

Eds.: K. O. Hodgson, B. Hedman, J. E. Penner-Hahn. Berlin – Heidelberg – New York – Tokyo: Springer 1984. XV, 533 S., 392 Abb., zahlr. Tab. u. Lit. (Springer Proc. Phys., Vol. 2); Hard cover DM 99,–

Es ist seit Jahren zur Gewohnheit geworden, daß internationale Tagungen ihre Materialien entweder als Sonderbände oder in laufender Reihe in einer einschlägigen Zeitschrift veröffentlichen. Dabei entstehen immer mehr Probleme des Umfangs und der Bearbeitungsdauer. Intervalle von einem Jahr zwischen Konferenz und Publikation sind selbst bei renommierten Verlagen keine Seltenheit mehr. Der Springer-Verlag bietet eine Alternative: Er publiziert die Kamera-fertigen Autorenmanuskripte von Tagungen mit physikalischen Themen innerhalb kürzester Frist in seiner neuen Reihe „Springer Proc. Phys.". Man wird sehen, wie weit dieses Konzept durchgehalten werden kann. Auf jeden Fall gibt es dafür gute Gründe, nicht zuletzt den der Entlastung der gängigen Zeitschriften mit ihren sich ohnehin ständig verlängernden Erscheinungsfristen.

Von den Herausgebern wurden die Beiträge der 3. internationalen EXAFS-Tagung wie folgt klassifiziert (jeweils in Klammern die Anzahl der betreffenden Arbeiten): Grundprobleme der Röntgenabsorption im kantenferneren (EXAFS) und kantennäheren Bereich (XANES) (20), Datenanalyse (7), biologische Systeme (16), katalytische Systeme und kleine Metallcluster (16), Oberflächenstrukturen (7), amorphe Materialien und Gläser (19), Geologie/Geochemie/Hochdruckprobleme (13), sonstige Anwendungen (23) und abschließend verwandte Techniken und Instrumentierung (18). Wie im Vorwort vermerkt, ist die Einteilung nicht willkürfrei, und im Interesse des (potentiellen) Lesers sei betont, daß die Einzelanwendungen stärker dominieren, als das in dieser Gliederung zum Ausdruck kommt. Rund die Hälfte der Publikationen des Grundlagen-Kapitels hätte man mit kaum schlechterer Begründung in die Anwendungen einreihen können, und die Herauslösung der Datenanalyse ist insgesamt nicht leicht zu rechtfertigen.

Der Tagungsinhalt lebt natürlich von den Möglichkeiten, welche die in neuerer Zeit erschlossene Anwendung der Synchrotronstrahlung zur Feinabtastung bei Anwendung der an sich schon viele Jahrzehnte bekannten Methode der Untersuchung von schwingungsartigen Überlagerungen auf den Absorptionsverläufen oberhalb der Kanten bietet. Für die Applikation ist es immer noch Standard, diese durch Interferenzen von auslaufender und an den Nachbaratomen teilweise rückgestreuter Photoelektronen-Welle zustande kommenden Verlaufsmuster mit Referenzmaterialien zu vergleichen, um Aussagen über Koordinationszahlen, Art der Liganden, mittlere Bindungsabstände und -richtungen sowie Schwankungsbreiten von Bildungslängen zu erhalten. Rein theoretisch-rechnerische Aussagen, etwa über den X_α-Streuwellenformalismus zur Strukturbeschreibung, sind möglich, stellen aber noch die seltene Ausnahme dar. Die Grundlagendiskussionen betreffen besonders den Komplex der Abgrenzung von Einzel- und Mehrfachstreuungen für die Interferenzen im Übergangs-Endzustand und die interessante Tatsache, daß dieser entscheidende Endzustand nicht nur aus dem allgemeinen Streuformalismus, sondern auch als Bloch-Zustand aus einer Bandstrukturberechnung mit periodischem Potential erhalten werden kann.

Die Vielzahl der Anwendungen bei niedriger Elementkonzentration (z. T. Spurenelemente, besonders im biologischen Bereich) macht zunehmend von den Möglichkeiten einer Messung in Emission Gebrauch, um die kleinen Effekte aus dem sonst riesigen Absorptionsuntergrund herauszuheben.

Als ausgesprochene Oberflächenmethode wird EXAFS brauchbar, wenn man sich solcher Signale bedient, die zwar insgesamt auf denselben interferenzbeeinflußten Übergängen beruhen, wie die kantennahe Photoabsorption selbst, aber aus dünner, oberflächennaher Schicht herrühren: *Auger*-, Photo- und Sekundärelektronen, Sekundärionen usw.. Die entsprechenden Experimente leiden zwar noch unter den geringen Signalgrößen und dem entsprechend schlechten Signal/Untergrund-Verhältnis und liefern vorläufig nur Bindungslängen und -richtungen, aber ihre Möglichkeiten und ihre Anzahl nehmen zu.

Mangels einer Synchrotron-Strahlenquelle sind in manchen Ländern nur wenige Fachleute auf den Gebieten von EXAFS und XANES tätig. Deren Interesse an diesem Tagungsband dürfte ausgesprochen groß sein, vermittelt er doch einen ausgezeichneten Überblick über den aktuellen Weltstand dieser Arbeitsrichtungen. *G. Brunner*

Otto Hahn
Ein Forscherleben unserer Zeit. 1879—1968
Große Naturforscher, Band 45

W. Gerlach, D. Hahn, Stuttgart: Wissenschaftl. Verlagsgesellschaft 1984, 265 Seiten, 53 Abb. (Große Naturforscher, Bd. 45); Kst. geb. DM 39,—

In den letzten Jahren sind anläßlich unterschiedlicher Jubiläen eine Vielzahl von historisch orientierten Veröffentlichungen und Biographien zur Radioaktivität, zur Entdeckung der Kernspaltung und zu den damit in Verbindung stehenden Forscherpersönlichkeiten: *Otto Hahn, Liese Meitner, Fritz Straßmann* etc. erschienen. Man sollte annehmen, daß diesen Publikationen kaum noch Wesentliches hinzugefügt werden kann.

Das vorliegende von *D. Hahn* herausgegebene und durch Zeittafeln ergänzte Buch wurde von dem bedeutenden und weithin bekannten deutschen Physiker *Walter Gerlach* — einem langjährigen Freund *Otto Hahns* — geschrieben.

Diese wissenschaftshistorische Arbeit *W. Gerlachs* erschien bereits 1969 in der Reihe „Abhandlungen und Berichte des Deutschen Museums, München". Obwohl literarisch ausgezeichnet, fand die Arbeit nur sehr geringe Verbreitung, so daß man sich entschloß, sie in der vorliegenden Form breiten Leserkreisen zugänglich zu machen. Ein solches Anliegen ist sehr zu begrüßen, da es sich um eine der bestgeschriebenen und authentischen Biographien *Otto Hahns* handelt. In sehr knapper und exakter Form werden hier die wichtigsten wissenschaftlichen und persönlichen Stationen dieses Forscherlebens von einem Kenner der Szene nachgezeichnet. So findet der Leser manches weniger Bekannte und Neue aus der privaten Sphäre der Familie Hahn, von der Internierung durch die Alliierten in Farmhall 1945/46, über das Denken und Wirken nach dem 2. Weltkrieg bis hin zum Todeslager in einer Göttinger Klinik.

Bei einer ersten oberflächlichen Betrachtung des Buches ist man überrascht, daß diese spannende Lebensgeschichte nur etwa die Hälfte des Textes ausmacht, dem sich im II. Teil umfangreiche Zeittafeln und bibliographische Zusammenstellungen anschließen. Bei näherer Betrachtung stellt sich jedoch bald heraus, daß diese an sich „trockenen" Zeittafeln und Aneinanderreihungen von Lehrveranstaltungen, Titeln, Ämtern, Auszeichnungen und Entdeckungen das Bild dieser außergewöhnlichen Forscherpersönlichkeit *Otto Hahn* vertiefen. Dazu tragen insbesondere die Zitate über *Otto Hahn*, die chronologische Tabelle seiner Entdeckungen, die nach Jahren geordnete Bibliographie aller seiner Veröffentlichungen sowie die geschickt ausgewählten Photographien bei.

Dieses Buch ist nicht nur historisch interessierten Lesern zu empfehlen; es wird auch den Studenten und jungen Wissenschaftlern zeigen, was wesentlich im Leben ist und worauf es in der wissenschaftlichen Arbeit ankommt.

H. Koch

Trace Element Analytical Chemistry in Medicine and Biology. Vol. 3
Proc. of the Third International Workshop, Neuherberg, F.R.G., April 1984

Eds.: P. Brätter, P. Schramel. Berlin — New York: Walter de Gruyter 1984. 761 S., 204 Abb., 127 Tab., 1258 Lit.; Hardcover DM 280.—

Die im Abstand von zwei Jahren in Neuherberg stattfindende Arbeitstagung wird jeweils von der Gesellschaft für Strahlenforschung und der Arbeitsgruppe „Spurenelemente in den Lebenswissenschaften" der Arbeitsgemeinschaft der Großforschungseinrichtungen (wie KFA Jülich, TH Darmstadt, Hahn-Meitner-Institut) getragen. Gegenstand des 3. Treffens war vor allem der Austausch neuerer Erkenntnisse auf dem Gebiet der essentiellen Spurenelemente Se, Zn und Mn (Analytik und Wirkmechanismus) sowie die Darstellung allgemeiner Probleme der Spurenelementanalytik in den Lebenswissenschaften. In 11 Sektionen wurden 11 Plenarvorträge, 14 Diskussionsbeiträge und 46 Poster gehalten.

Folgende Themenkomplexe wurden im einzelnen behandelt: Se-Metabolismen; Selen in der Ernährung; Selenanalyse; Mangan in Ernährung und Medizin; Wirkformen von Spurenelementen; Rolle der Spurenelemente bei cardiovasculären Erkrankungen; Anwendung verschiedener analytischer Techniken; Neuere essentielle Spurenelemente; Analyse von Spurenelementen in biologischen Materialien; Standardreferenzmaterialien in der biomedizinischen Spurenelementanalytik; Bedeutung von Spurenelementen in der Humanmedizin.

Die Vielzahl der Einzelinformationen und die Breite der vorgetragenen neueren Erkenntnisse schließt eine Aufzählung an dieser Stelle aus. Folgendes nur soll hervorgehoben werden:

- das fortentwickelte analytische Instrumentarium zur Bewältigung dieser schwierigen analytischen Fragestellungen. Neben den üblichen Standardverfahren, wie Photometrie, Voltametrie, Aktivierungsanalyse und Atomabsorptionsspektroskopie in allen bekannten Varianten sind es auch solche Methoden, die in der Festkörperanalytik ihren Platz gefunden haben wie die Röntgenfluoreszenzanalyse, auch gekoppelt mit Scanning-Elektronenmikroskopie (Partikelanalyse in menschlichem Lungengewebe), die PIXE und die Scanning-PIXE (Spurenelementbestimmungen in Mikrobereichen, Aufnahme von Multielementkonzentrationsprofilen an Nierengewebe).
- die Kombination verschiedener analytischer Methoden, die Sorgfalt bei der Probenpräparation und die umfassende statistische Bewertung der erhaltenen Ergebnisse. Als Lehrstück für eine umfassende analytische Problemaufbereitung und stellvertretend für viele andere soll der Plenarvortrag von Prof. *G. Tölg* vom Max-Planck-Institut für Metallforschung, Stuttgart, angeführt werden, der über den Stand der Se-Analyse in biologischem Material berichtete und neben den eigentlichen Bestimmungsmethoden den Gesamtkomplex der Probenaufbereitung und der Fehlerquellen ausführlich diskutierte.
- die Zuwendung der großen Kernforschungszentren der BRD und Westeuropas zu diesem neuen Arbeitsgebiet und die umfassenden Aktivitäten der IAEA auf diesem Gebiet.

Wie sehr der Problemkreis „Spurenelementanalytik in den Lebenswissenschaften" in Fluß geraten ist, zeigt die Analyse der zitierten Literatur. Von den 1258 Literaturstellen enstammt über die Hälfte dem Zeitraum 1981—85, nur 98 Zitate sind älter als 15 Jahre und das bei 11 Plenarvorträgen. Bemerkenswert ist die weitverzweigte interdisziplinäre Kooperation der verschiedenen Forschungsgruppen.

Die Lektüre dieses Proceedings-Bandes sowie seiner zwei Vorläufer sei all denen angeraten, die sich mit dem Problem der Spurenelementanalytik auseinandersetzen oder analytische Aufgabenstellung lösen müssen, unabhängig vom Einzugsbereich, der Aufgabenstellung und der Matrix.

J. Flachowsky

Berichtigung

Heft 7/1985, S. 257: Statt *S. Rieedl* (Autorenname unter dem Beitrag „Aus der Praxis") lies *S. Riedel.*

Journal of Information Recording Materials
Journal für Signalaufzeichnungsmaterialien

Herausgegeben von dem VEB Fotochemisches Kombinat Wolfen

Die Zeitschrift veröffentlicht Beiträge in Englisch, Deutsch und Russisch; sie widmet sich sowohl theoretischen als auch anwendungstechnischen Problemen der wissenschaftlichen Photographie, der magnetischen und anderer Nichthalogensilberaufzeichnungsmaterialien sowie -verfahren. Zum Profil der Zeitschrift gehören ebenso Abhandlungen über informationstheoretische Grundlagen wie Beiträge zur Herstellung moderner Informationsaufzeichnungsmaterialien.

Interessenten sind Wissenschaftler und Anwendungstechniker, die in Industrie, Forschung und Lehre auf dem Gebiet der Physik, Chemie, Informationsverarbeitung bis hin zur Mikroelektronik tätig sind; darüber hinaus wissenschaftliche Bibliotheken, physikalisch-technische sowie physikalisch-chemische Institute.

Erscheinungsweise und Bezugsbedingungen:

Die Zeitschrift erscheint zweimonatlich. Die jährlich erscheinenden 6 Hefte bilden einen Band.
Format: 16,7 cm × 24,0 cm.
Jährlicher Bezugspreis: 120,— DM zuzüglich Versandspesen. Einzelheft: 20,— DM.
ISSN 0323-598X Bestellnummer der Zeitschrift: 1089.
1985 erscheint Band 13.

Bezugsmöglichkeiten:

Bestellungen sind zu richten

- in der DDR: an den Postzeitungsvertrieb unter Angabe der Kundennummer des Bestellenden oder an den AKADEMIE-VERLAG BERLIN, DDR-1086 Berlin, Leipziger Str. 3—4;
- im sozialistischen Ausland: an eine Buchhandlung für fremdsprachige Literatur oder an den zuständigen Postzeitungsvertrieb;
- in der BRD und Berlin(West): an eine Buchhandlung oder an die Auslieferungsstelle KUNST UND WISSEN, Erich Bieber OHG, Wilhelmstr. 4—6, D-7000 Stuttgart 1;
- in den übrigen westeuropäischen Ländern: an eine Buchhandlung oder an die Auslieferungsstelle KUNST UND WISSEN, Erich Bieber GmbH, Dufourstr. 51, CH-8008 Zürich;
- im übrigen Ausland an den Internationalen Buch- und Zeitschriftenhandel; den Buchexport, Volkseigener Außenhandelsbetrieb der DDR, DDR-7010 Leipzig, Postfach 160; oder an den AKADEMIE-VERLAG BERLIN, DDR-1086 Berlin, Leipziger Straße 3—4.

AKADEMIE-VERLAG BERLIN
1086 Berlin, Leipziger Straße 3—4

ISOTOPENPRAXIS

Band 21 | 1985 | Heft 12

Contents | Inhalt

▶**Language of Publication** ▶**Publikationssprache**

Isotopenpraxis is indexed in Current Contents/PC & ES, Science Citation Index, and INIS Atomindex

MIX
Papier | Fördert gute Waldnutzung
FSC® C083411

Zeitfracht Medien GmbH
Ferdinand-Jühlke-Straße 7
99095 Erfurt, Deutschland
produktsicherheit@kolibri360.de